T0143123

Quantum Attacks on Public-Key Cryptosystems

Song Y. Yan

Quantum Attacks on Public-Key Cryptosystems

Springer

Song Y. Yan
Department of Mathematics
Harvard University
Cambridge, MA
USA

ISBN 978-1-4899-9779-1   ISBN 978-1-4419-7722-9 (eBook)
DOI 10.1007/978-1-4419-7722-9
Springer New York Heidelberg Dordrecht London

Table of Contents

Preface

If we knew what it was we were doing, it would not be called research, would it?

ALBERT EINSTEIN (1879–1955)
The 1921 Nobel Laureate in Physics

In research, if you know what you are doing, then you shouldn't be doing it.

RICHARD HAMMING (1915–1998)
The 1968 Turing Award Recipient

It is well known that the security of the most widely used public-key cryptosystems such as RSA (Rivest-Shamir-Adleman), DSA (digital signature algorithm), and ECC (elliptic curve cryptography) relies on the intractability of one of the following three number-theoretic problems, namely, the integer factorization problem (IFP), the discrete logarithm problem (DLP), and the elliptic curve discrete logarithm problem (ECDLP). Since no polynomial-time algorithms have been found so far for solving these three hard problems, the cryptosystems based on them are secure. There are, however, quantum algorithms, due to Shor and others, which can solve these three intractable problems in polynomial time, provided that a practical quantum computer can be constructed.

The monograph provides a quantum approach to solve all these three intractable number-theoretic problems and to attack the cryptosystems based on these three problems. The organization of the book is as follows. Chapter 1 provides an introduction to the basic concepts and ideas of quantum computation. Chapter 2 discusses shor's quantum factoring algorithm and its application to the cryptanalysis of IPF-based, particularly RSA cryptosystems. Chapter 2 discusses Shor's quantum discrete logarithm algorithm and its application to the cryptanalysis of DLP-based cryptosystems.

Chapter 4 is devoted to the study of the extension of Shor's quantum algorithms for solving the ECDLP problems and the attacks on the ECDLP-based cryptosystems. Finally in Chapter 5, some quantum resistant public-key cryptosystems are studied, which can be used in the post-quantum age.

The monograph is a revised and extended version of the author's earlier version *Cryptanalytic Attacks on RSA*, with an emphasis on quantum attacks for public-key cryptography. It is self-contained and can be used as a basic reference for computer scientists, mathematicians, electrical engineers, and physicists, interested in quantum computation and quantum cryptography. It can also be used as a final year undergraduate or a 1st-year graduate text in the field.

Acknowledgments

The author would like to thank the three anonymous referees for their very helpful suggestions and comments. Special thanks must be given to Prof Michael Sipser and Prof Ronald Rivest at MIT, Prof Benedict Gross at Harvard, Susan Lagerstrom-Fife, Courtney Clark and Jennifer Maurer at Springer New York, for their encouragement, support, and help. The research was supported in part by the Royal Academy of Engineering, London, the Royal Society, London, Harvard University, Massachusetts Institute of Technology, and Wuhan University.

Finally, the author would specifically like to thank Prof Yanxiang He, Dean of Computer School of Wuhan University for his encouragement, support, and collaboration.

Cambridge, MA   S.Y. Yan

1. Classic and Quantum Computation

Anyone who is not shocked by quantum theory has not understood it.

NIELS BOHR (1885–1962)
The 1922 Nobel Laureate in Physics

In this chapter, we shall first give an account of the basic concepts and results in classical computability and complexity and then, the quantum computability and complexity, which will be used throughout the book.

1.1 Classical Computability Theory

Computability studies what a computer can do and what a computer cannot do. As a Turing machine can do everything that a real computer can do, our study of computability will be within the theoretical framework of Turing machines.

Turing Machines

The idea and the theory of Turing machines were first proposed and studied by the great English logician and mathematician Alan Turing (1912–1954) in his seminal paper [43] published in 1936 (see Fig. 1.1). First of all, we shall present a formal definition of the Turing machine.

Definition 1.1. A standard multitape *Turing machine*, M (see Fig. 1.2), is an algebraic system defined by

$$M = (Q, \Sigma, \Gamma, \delta, q_0, \square, F) \tag{1.1}$$

where

S.Y. Yan, *Quantum Attacks on Public-Key Cryptosystems*,
DOI 10.1007/978-1-4419-7722-9_1,
© Springer Science+Business Media, LLC 2013

230   A. M. Turing   [Nov. 12,

ON COMPUTABLE NUMBERS, WITH AN APPLICATION TO
THE ENTSCHEIDUNGSPROBLEM

By A. M. Turing.

[Received 28 May, 1936.—Read 12 November, 1936.]

The "computable" numbers may be described briefly as the real numbers whose expressions as a decimal are calculable by finite means. Although the subject of this paper is ostensibly the computable *numbers*, it is almost equally easy to define and investigate computable functions of an integral variable or a real or computable variable, computable predicates, and so forth. The fundamental problems involved are, however, the same in each case, and I have chosen the computable numbers for explicit treatment as involving the least cumbrous technique. I hope shortly to give an account of the relations of the computable numbers, functions, and so forth to one another. This will include a development of the theory of functions of a real variable expressed in terms of computable numbers. According to my definition, a number is computable if its decimal can be written down by a machine.

In §§ 9, 10 I give some arguments with the intention of showing that the computable numbers include all numbers which could naturally be regarded as computable. In particular, I show that certain large classes of numbers are computable. They include, for instance, the real parts of all algebraic numbers, the real parts of the zeros of the Bessel functions, the numbers π, e, etc. The computable numbers do not, however, include all definable numbers, and an example is given of a definable number which is not computable.

Although the class of computable numbers is so great, and in many ways similar to the class of real numbers, it is nevertheless enumerable. In § 8 I examine certain arguments which would seem to prove the contrary. By the correct application of one of these arguments, conclusions are reached which are superficially similar to those of Gödel†. These results

† Gödel, "Über formal unentscheidbare Sätze der Principia Mathematica und verwandter Systeme, I", *Monatshefte Math. Phys.*, 38 (1931), 173–198.

Figure 1.1. Alan Turing and the first page of his 1936 paper

1. Q is a finite set of *internal states*.
2. Σ is a finite set of symbols called the *input alphabet*. We assume that $\Sigma \subseteq \Gamma - \{\square\}$.
3. Γ is a finite set of symbols called the *tape alphabet*.
4. δ is the transition function, which is defined by
   (a) If M is a deterministic Turing machine (DTM), then
   $$\delta: Q \times \Gamma^k \to Q \times \Gamma^k \times \{L, R\}^k \qquad (1.2)$$
   (b) If M is a nondeterministic Turing machine (NDTM), then
   $$\delta: Q \times \Gamma^k \to 2^{Q \times \Gamma^k \times \{L,R\}^k} \qquad (1.3)$$
   where L and R specify the movement of the read–write head *left* or *right*. When $k = 1$, it is just a standard one-tape Turing machine.
5. $\square \in \Gamma$ is a special symbol called the *blank*.
6. $q_0 \in Q$ is the *initial state*.
7. $F \subseteq Q$ is the set of *final states*.

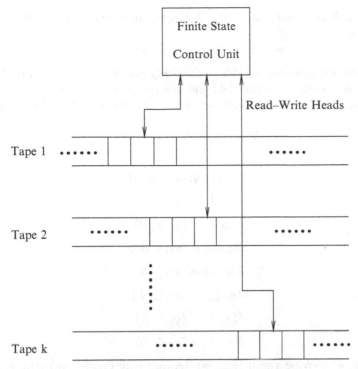

Figure 1.2. k-tape $(k \geqslant 1)$ Turing machine

Turing machines, although simple and abstract, provide us with a most suitable model of computation for modern *digital* and even *quantum* computers.

Example 1.1. Given two positive integers x and y, design a Turing machine that computes $x+y$. First, we have to choose some convention for representing positive integers. For simplicity, we will use unary notation in which any positive integer x is represented by $w(x) \in \{1\}^+$, such that $|w(x)| = x$. Thus in this notation, 4 will be represented by 1111. We must also decide how x and y are placed on the tape initially and how their sum is to appear at the end of the computation. It is assumed that $w(x)$ and $w(y)$ are on the tape in unary notation, separated by a single 0, with the read–write head on the leftmost symbol of $w(x)$. After the computation, $w(x + y)$ will be on the tape followed by a single 0, and the read–write head will be positioned at the left end of the result. We therefore want to design a Turing machine for performing the computation

$$q_0 w(x) 0 w(y) \overset{*}{\vdash} q_f w(x + y) 0,$$

where $q_f \in F$ is a final state and $\overset{*}{\vdash}$ indicates an unspecified number of steps as follows:

$$q_0 w(x) 0 w(y) \vdash \cdots \vdash q_f w(x+y) 0.$$

Constructing a program for this is relatively simple. All we need to do is to move the separating 0 to the right end of $w(y)$, so that the addition amounts to nothing more than the coalition of the two strings. To achieve this, we construct

$$M = (Q, \Sigma, \Gamma, \delta, q_0, \square, F),$$

with

$$Q = \{q_0, q_1, q_2, q_3, q_4\},$$

$$F = \{q_4\},$$

$$\delta(q_0, 1) = (q_0, 1, R),$$

$$\delta(q_0, 0) = (q_1, 1, R),$$

$$\delta(q_1, 1) = (q_1, 1, R),$$

$$\delta(q_1, \square) = (q_2, \square, L),$$

$$\delta(q_2, 1) = (q_3, 0, L),$$

$$\delta(q_3, 1) = (q_3, 1, L),$$

Note that in moving the 0 right we temporarily create an extra 1, a fact that is remembered by putting the machine into state q_1. The transition $\delta(q_2, 1) = (q_3, 0, R)$ is needed to remove this at the end of the computation. This can be seen from the sequence of instantaneous descriptions for adding 111 to 11:

$$
\begin{aligned}
q_0 1110011 \;\; &\vdash \;\; 1 q_0 110011 \\
&\vdash \;\; 11 q_0 1011 \\
&\vdash \;\; 111 q_0 011 \\
&\vdash \;\; 1111 q_1 11 \\
&\vdash \;\; 11111 q_1 1 \\
&\vdash \;\; 111111 q_1 \\
&\vdash \;\; 11111 q_2 1 \\
&\vdash \;\; 1111 q_3 10 \\
&\;\;\;\vdots \\
&\vdash \;\; q_3 \square 111110 \\
&\vdash \;\; q_4 111110,
\end{aligned}
$$

or briefly as follows:

$$q_0 1110011 \overset{*}{\vdash} q_4 111110.$$

The Church–Turing Thesis

Any *effectively* computable function can be computed by a Turing machine, and there is no effective procedure that a Turing machine cannot perform. This leads naturally to the following famous Church–Turing thesis, named after Alonzo Church and Alan Turing:

> **The Church–Turing Thesis.** Any effectively computable function can be computed by a Turing machine.

The Church–Turing thesis thus provides us with a powerful tool to distinguish what is computation and what is not computation, what function is computable and what function is not computable, and more generally, what computers can do and what computers cannot do.

It must be noted that the Church–Turing thesis is not a mathematical theorem, and hence it cannot be proved formally, since to prove the Church–Turing thesis, we need to formalize what is effectively computable, which is impossible. However, many computational evidences support the thesis, and in fact no counterexample has been found yet.

Remark 1.1. Church in his famous 1936 paper [7] (the first page of the paper can be found in Fig. 1.3) proposed the important concept of λ-definable, and later in his book review [8] on Turing's 1936 paper, he said that *all effective procedures are in fact Turing equivalent*. This is what we call now the Church–Turing thesis. It is interesting to note that Church was the PhD advisor of Alan Turing (1938), Michael Rabin (1957), and Dana Scott (1958), all at Princeton; Rabin and Scott were also the 1976 Turing Award Recipients, a prize considered as an equivalent Nobel Prize in Computer Science.

Decidability and Computability

Although a Turing machine can do everything that a real computer can do, there are, however, many problems that Turing machines cannot do; the simplest is actually related to the Turing machine itself, the so-called *Turing machine halting problem*.

Definition 1.2. A language is *Turing-acceptable* if there exists a Turing machine that accepts the language. A Turing-acceptable language is also called a *recursively enumerable language*.

When a Turing machine starts on an input, there are three possible outcomes: accept, reject, or loop (i.e., the machine falls into an infinite loop without any output). If a machine can always make a decision to accept or reject a language, then the machine is said to decide the language.

Definition 1.3. A language is *Turing-decidable* if there exists a Turing machine that decides the language. A Turing-decidable language is also called *recursive language*.

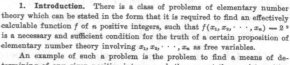

AN UNSOLVABLE PROBLEM OF ELEMENTARY NUMBER
THEORY.[1]

By ALONZO CHURCH.

1. Introduction. There is a class of problems of elementary number theory which can be stated in the form that it is required to find an effectively calculable function f of n positive integers, such that $f(x_1, x_2, \cdots, x_n) = 2$ [2] is a necessary and sufficient condition for the truth of a certain proposition of elementary number theory involving $x_1, x_2, \cdots, x_n$ as free variables.

An example of such a problem is the problem to find a means of determining of any given positive integer n whether or not there exist positive integers x, y, z, such that $x^n + y^n = z^n$. For this may be interpreted, required to find an effectively calculable function f, such that $f(n)$ is equal to 2 if and only if there exist positive integers x, y, z, such that $x^n + y^n = z^n$. Clearly the condition that the function f be effectively calculable is an essential part of the problem, since without it the problem becomes trivial.

Another example of a problem of this class is, for instance, the problem of topology, to find a complete set of effectively calculable invariants of closed three-dimensional simplicial manifolds under homeomorphisms. This problem can be interpreted as a problem of elementary number theory in view of the fact that topological complexes are representable by matrices of incidence. In fact, as is well known, the property of a set of incidence matrices that it represent a closed three-dimensional manifold, and the property of two sets of incidence matrices that they represent homeomorphic complexes, can both be described in purely number-theoretic terms. If we enumerate, in a straightforward way, the sets of incidence matrices which represent closed three-dimensional manifolds, it will then be immediately provable that the problem under consideration (to find a complete set of effectively calculable invariants of closed three-dimensional manifolds) is equivalent to the problem, to find an effectively calculable function f of positive integers, such that $f(m, n)$ is equal to 2 if and only if the m-th set of incidence matrices and the n-th set of incidence matrices in the enumeration represent homeomorphic complexes.

Other examples will readily occur to the reader.

[1] Presented to the American Mathematical Society, April 19, 1935.
[2] The selection of the particular positive integer 2 instead of some other is, of course, accidental and non-essential.

345

Figure 1.3. Alonzo Church and the first page of his 1936 paper

Definition 1.4. The Turing machine halting problem may be defined as follows:

$$L_{\text{TM}} = \{(M, w) \mid M \text{ is a Turing machine and } M \text{ accepts } w\}.$$

Theorem 1.1. L_{TM} is undecidable.

Turing machines that always halt are good model of an *algorithm*, a well-defined sequence of steps that always finishes and produces an answer. If an algorithm for a given problem exists, then the problem is *decidable*. Let the language L be a *problem*; then L is decidable if it is recursive language, and it is undecidable if it is not recursive language. From a practical point of view, the existence or nonexistence of an algorithm to solve a problem is of more important than the existence or nonexistence of a Turing machine to solve the problem. So, to distinguish problems or languages between decidable or undecidable is of more important than that between recursively enumerable and non-recursively enumerable. Figure 1.4 shows the relationship among the three classes of problems/languages.

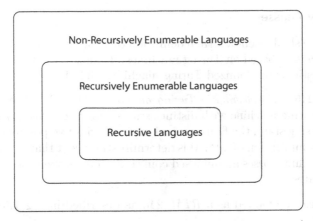

Figure 1.4. Relationships among recursive-related languages/problems

Exercises and Problems for Sect. 1.1

1. Explain why a Turing machine can do everything that a real computer can do.

2. Explain why Church–Turing thesis cannot be proved rigorously.

3. Explain why all different types of Turing machines such as single-tape Turing machines and multiple-tape Turing machines are equivalent.

4. Show that there is a language that is recursively enumerable but not recursive.

5. Show that the Turing machine halting problem is undecidable.

6. Give one more example (problem) that is undecidable.

1.2 Classical Complexity Theory

Computability is only concerned with what computer can do, but ignores the the computing resources such as the time and space required for completing a computation task. Computational complexity, on the other hand, fills this gap by considering mainly the computing resources such as the time and space required for completing a computation task. Thus, a theoretically computable problem may be practically uncomputable if it required too much time such as 50 million years or too much space. In this section, we shall study mainly the time complexity of computational problems.

Complexity Classes

First of all, we shall resent a formal definition of some common computational complexity classes based on Turing machines. To do so, we need a definition for probabilistic or randomized Turing machines (RTMs).

Definition 1.5. A *probabilistic Turing machine* (PTM) is a type of nondeterministic Turing machine with distinct states called *coin-tossing states*. For each coin-tossing state, the finite control unit specifies two possible legal next states. The computation of a PTM is deterministic except that in coin-tossing states the machine tosses an unbiased coin to decide between the two *possible legal* next states.

A PTM can be viewed as a *RTM* [24], as described in Fig. 1.5. The first

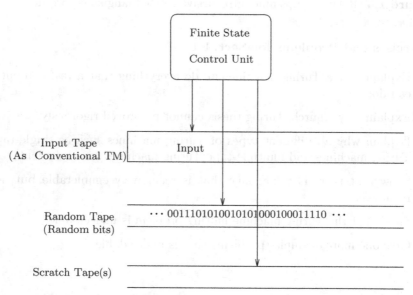

Figure 1.5. Randomized Turing machine

tape, holding input, is just the same as conventional multitape Turing machine. The second tape is referred to as random tape, containing randomly and independently chosen bits, with probability $1/2$ of a 0 and the same probability $1/2$ of a 1. The third and subsequent tapes are used, if needed, as scratch tapes by the Turing machine.

Definition 1.6. $\mathcal{P}$ is the class of problems solvable in polynomial time by a DTM. Problems in this class are classified to be tractable (feasible) and easy to solve on a computer. For example, additions of any two integers, no matter how big they are, can be performed in polynomial time, and hence it is in $\mathcal{P}$.

Definition 1.7. $\mathcal{NP}$ is the class of problems solvable in polynomial time on a NDTM. Problems in this class are classified to be intractable (infeasible) and hard to solve on a computer. For example, the traveling salesman problem (TSP) is in $\mathcal{NP}$, and hence it is hard to solve.

In terms of formal languages, we may also say that $\mathcal{P}$ is the class of languages where the membership in the class can be decided in polynomial time, whereas $\mathcal{NP}$ is the class of languages where the membership in the class can be verified in polynomial time [41]. It seems that the power of polynomial-time verifiable is greater than that of polynomial-time decidable, but no proof has been given to support this statement (see Fig. 1.6). The question of whether or not $\mathcal{P} = \mathcal{NP}$ is one of the greatest unsolved problems in computer science and mathematics, and in fact it is one of the seven Millennium Prize Problems proposed by the Clay Mathematics Institute in Boston in 2000, each with one-million US dollars [12].

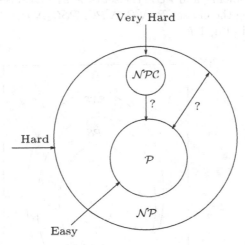

Figure 1.6. The $\mathcal{P}$ versus $\mathcal{NP}$ problem

Definition 1.8. $\mathcal{EXP}$ is the class of problems solvable by a DTM in time bounded by 2^{n^i}.

Definition 1.9. A function f is polynomial-time computable if for any input w, $f(w)$ will halt on a Turing machine in polynomial time. A language A is polynomial-time reducible to a language B, denoted by $A \leqslant_P B$, if there exists a polynomial-time computable function such that for every input w,

$$w \in A \iff f(w) \in B.$$

The function f is called the polynomial-time reduction of A to B.

Definition 1.10. A language/problem L is $\mathcal{NP}$-Completeness if it satisfies the following two conditions:

1. $L \in \mathcal{NP}$,

2. $\forall A \in \mathcal{NP}, \ A \leqslant_{\mathcal{P}} L$.

Definition 1.11. A problem D is $\mathcal{NP}$-hard if it satisfies the following condition:

$$\forall A \in \mathcal{NP}, \ \ A \leqslant_{\mathcal{P}} D$$

where D may be in $\mathcal{NP}$ or may not be in $\mathcal{NP}$. Thus, $\mathcal{NP}$-hard means *at least as hard as any $\mathcal{NP}$-problem*, although it might, in fact, be harder.

Similarly, one can define the class of problems of $\mathcal{P}$-SPACE, $\mathcal{P}$-SPACE-complete, and $\mathcal{P}$-SPACE-hard. We shall use $\mathcal{NPC}$ to denote the set of $\mathcal{NP}$-complete problems, $\mathcal{PSC}$ the set of $\mathcal{P}$-SPACE-complete problems, $\mathcal{NPH}$ the set of $\mathcal{NP}$-hard problems, and $\mathcal{PSH}$ the set of $\mathcal{P}$-SPACE-hard problems. The relationships among the classes $\mathcal{P}, \mathcal{NP}, \mathcal{NPC}, \mathcal{PSC}, \mathcal{NPH}, \mathcal{PSH}$, and $\mathcal{EXP}$ may be described in Fig. 1.7.

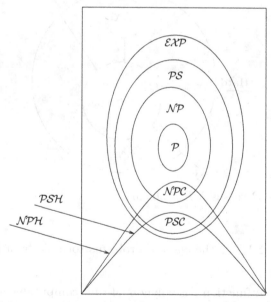

Figure 1.7. Conjectured relationships among classes $\mathcal{P}, \mathcal{NP}$ and $\mathcal{NPC}$, etc.

Definition 1.12. $\mathcal{RP}$ is the class of problems solvable in expected polynomial time with *one-sided error* by a probabilistic (randomized) Turing machine (PTM). By "one-sided error" we mean that the machine will answer "yes" when the answer is "yes" with a probability of error $< 1/2$ and will answer "no" when the answer is "no" with zero probability of error.

Definition 1.13. $\mathcal{ZPP}$ is the class of problems solvable in expected polynomial time with *zero error* on a PTM. It is defined by $\mathcal{ZPP} = \mathcal{RP} \cap$ co-$\mathcal{RP}$, where co-$\mathcal{RP}$ is the complement of $\mathcal{RP}$, or co-$\mathcal{RP}$ is the complementary language of $\mathcal{RP}$, i.e., co-$\mathcal{RP} = \{L : \overline{L} \in \mathcal{RP}\}$). By "zero error" we mean that the machine will answer "yes" when the answer is "yes" (with zero probability of error) and will answer "no" when the answer is "no" (also with zero probability of error). But note that the machine may also answer "?," which means that the machine does not know whether the answer is "yes" or "no." However, it is guaranteed that at most half of simulation cases, the machine will answer "?." $\mathcal{ZPP}$ is usually referred to an *elite class*, because it also equals to the class of problems that can be solved by randomized algorithms that always give the correct answer and run in expected polynomial time.

Definition 1.14. $\mathcal{BPP}$ is the class of problems solvable in expected polynomial time with *two-sided error* on a PTM, in which the answer always has probability at least $\frac{1}{2} + \delta$, for some fixed $\delta > 0$ of being correct. The "$\mathcal{B}$" in $\mathcal{BPP}$ stands for "bounded away the error probability from $\frac{1}{2}$"; for example, the error probability could be $\frac{1}{3}$.

The space complexity classes $\mathcal{P}$-SPACE and $\mathcal{NP}$-SPACE can be defined analogously as $\mathcal{P}$ and $\mathcal{NP}$. It is clear that a time class is included in the corresponding space class since one unit is needed to the space by one square. Although it is not known whether or not $\mathcal{P} = \mathcal{NP}$, it is known that $\mathcal{P}$-SPACE $= \mathcal{NP}$-SPACE. It is generally believed that

$$\mathcal{P} \subseteq \mathcal{ZPP} \subseteq \mathcal{RP} \subseteq \left(\begin{array}{c} \mathcal{BPP} \\ \mathcal{NP} \end{array} \right) \subseteq \mathcal{P}\text{-SPACE} \subseteq \mathcal{EXP}.$$

Besides the proper inclusion $\mathcal{P} \subset \mathcal{EXP}$, it is not known whether any of the other inclusions in the above hierarchy is proper. Note that the relationship of $\mathcal{BPP}$ and $\mathcal{NP}$ is not known, although it is believed that $\mathcal{NP} \nsubseteq \mathcal{BPP}$. Figure 1.8 shows the relationships among the various common complexity classes.

The Cook–Karp Thesis

It is widely believed, although no proof has been given, that problems in $\mathcal{P}$ are computationally tractable (or feasible, easy), whereas problems not in (i.e., beyond) $\mathcal{P}$ are computationally intractable (or infeasible, hard, difficult). This is the famous *Cook–Karp thesis*, named after Stephen Cook, who first studied the $\mathcal{P}$–$\mathcal{NP}$ problem (the first page of Cook's paper can be found in Fig. 1.9) and Richard Karp, who proposed a list of the $\mathcal{NP}$-complete problems (the first page of Karp's paper can be found in Fig. 1.10).

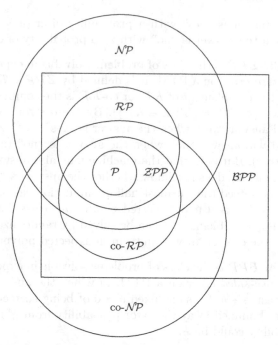

Figure 1.8. Conjectured relationships among some common complexity classes

The Cook–Karp thesis. Any computationally tractable problem can be computed by a Turing machine in deterministic polynomial time.

Thus, problems in $\mathcal{P}$ are tractable whereas problems in $\mathcal{NP}$ are intractable. However, there is not a clear cut between the two types of problems. This is exactly the hard $\mathcal{P}$ versus $\mathcal{NP}$ problem, mentioned earlier. Compared to the Church–Turing thesis, the Cook–Karp thesis provides a step closer to practical computability and complexity, and hence, the life after Cook and Karp is much easier, since there is no need to go all the way back to Church and Turing. Again, Cook–Karp thesis is not a mathematical theorem and hence cannot be proved mathematically; however, evidences support the thesis.

Exercises and Problems for Sect. 1.2

1. Define and explain the following complexity classes:

$\mathcal{P}$,
$\mathcal{NP}$,

The Complexity of Theorem-Proving Procedures

Stephen A. Cook

University of Toronto

Summary

It is shown that any recognition problem solved by a polynomial time-bounded nondeterministic Turing machine can be "reduced" to the problem of determining whether a given propositional formula is a tautology. Here "reduced" means, roughly speaking, that the first problem can be solved deterministically in polynomial time provided an oracle is available for solving the second. From this notion of reducible, polynomial degrees of difficulty are defined, and it is shown that the problem of determining tautologyhood has the same polynomial degree as the problem of determining whether the first of two given graphs is isomorphic to a subgraph of the second. Other examples are discussed. A method of measuring the complexity of proof procedures for the predicate calculus is introduced and discussed.

Throughout this paper, a set of strings means a set of strings on some fixed, large, finite alphabet Σ. This alphabet is large enough to include symbols for all sets described here. All Turing machines are deterministic recognition devices, unless the contrary is explicitly stated.

1. Tautologies and Polynomial Re-Reducibility.

Let us fix a formalism for the propositional calculus in which formulas are written as strings on Σ. Since we will require infinitely many proposition symbols (atoms), each such symbol will consist of a member of Σ followed by a number in binary notation to distinguish that symbol. Thus a formula of length n can only have about n/logn distinct function and predicate symbols. The logical connectives are & (and), ∨ (or), and ⌐(not).

The set of tautologies (denoted by {tautologies}) is a

certain recursive set of strings on this alphabet, and we are interested in the problem of finding a good lower bound on its possible recognition times. We provide no such lower bound here, but theorem 1 will give evidence that {tautologies} is a difficult set to recognize, since many apparently difficult problems can be reduced to determining tautologyhood. By reduced we mean, roughly speaking, that if tautologyhood could be decided instantly (by an "oracle") then these problems could be decided in polynomial time. In order to make this notion precise, we introduce query machines, which are like Turing machines with oracles in [1].

A query machine is a multitape Turing machine with a distinguished tape called the query tape, and three distinguished states called the query state, yes state, and no state, respectively. If M is a query machine and T is a set of strings, then a T-computation of M is a computation of M in which initially M is in the initial state and has an input string w on its input tape, and each time M assumes the query state there is a string u on the query tape, and the next state M assumes is the yes state if u∈T and the no state if u∉T. We think of an "oracle", which knows T, placing M in the yes state or no state.

Definition

A set S of strings is P-reducible (P for polynomial) to a set T of strings iff there is some query machine M and a polynomial Q(n) such that for each input string w, the T-computation of M with input w halts within Q(|w|) steps (|w| is the length of w) and ends in an accepting state iff w∈S.

It is not hard to see that P-reducibility is a transitive relation. Thus the relation B on

Figure 1.9. Stephen Cook and the first page of his 1971 paper

$\mathcal{RP}$,

$\mathcal{BPP}$,

$\mathcal{ZPP}$,

$\mathcal{NP}$-complete,

$\mathcal{NP}$-hard,

$\mathcal{P}^{\#\mathcal{P}}$,

$\mathcal{P}$-SPACE,

$\mathcal{NP}$-SPACE, and

$\mathcal{EXP}$.

2. Show that $\mathcal{P} \subset \mathcal{RP}$.

3. Let SAT denote the *satisfiability problem*. Show that

$$\text{SAT} \in \mathcal{NP},$$

REDUCIBILITY AMONG COMBINATORIAL PROBLEMS[+]

Richard M. Karp

University of California at Berkeley

Abstract: A large class of computational problems involve the determination of properties of graphs, digraphs, integers, arrays of integers, finite families of finite sets, boolean formulas and elements of other countable domains. Through simple encodings from such domains into the set of words over a finite alphabet these problems can be converted into language recognition problems, and we can inquire into their computational complexity. It is reasonable to consider such a problem satisfactorily solved when an algorithm for its solution is found which terminates within a number of steps bounded by a polynomial in the length of the input. We show that a large number of classic unsolved problems of covering, matching, packing, routing, assignment and sequencing are equivalent, in the sense that either each of them possesses a polynomial-bounded algorithm or none of them does.

1. INTRODUCTION

All the general methods presently known for computing the chromatic number of a graph, deciding whether a graph has a Hamilton circuit, or solving a system of linear inequalities in which the variables are constrained to be 0 or 1, require a combinatorial search for which the worst case time requirement grows exponentially with the length of the input. In this paper we give theorems which strongly suggest, but do not imply, that these problems, as well as many others, will remain intractable perpetually.

[+]This research was partially supported by National Science Foundation Grant GJ-474.

85

Figure 1.10. Richard Karp and the first page of his 1972 paper

and
$$\text{SAT} \in \mathcal{NP}-\text{complete}.$$

4. Let HPP denote the *Hamiltonian path problem*. Show that
$$\text{HPP} \in \mathcal{NP},$$
and
$$\text{HPP} \in \mathcal{NP}-\text{complete}.$$

5. Show that HPP is polynomial-time reducible to TSP.

6. Prove or disprove $\mathcal{P} \neq \mathcal{NP}$.

7. Just the same as that it is not known if $\mathcal{P} \neq \mathcal{NP}$, it is also currently not known if $\mathcal{BPP} \neq \mathcal{P}\text{-SPACE}$, and proving or disproving this would be a major breakthrough in computational complexity theory. Prove or disprove
$$\mathcal{BPP} \neq \mathcal{P}\text{-SPACE}.$$

1.3 Quantum Information and Computation

The idea that computers can be viewed as physical objects and computations as physical processes is revolutionary; It was conceived by several scientists, most notably Richard Feynman (1918–1988) and David Deutsch (Born 1953). For example, Feynman published posthumously a book *Feynman Lectures on Computation* [17] (see Fig. 1.11) in 1996, where he introduced the theory of

Figure 1.11. Richard Feynman and the cover of his book

reversible computation, quantum mechanical computers, and quantum aspects of computation in great detail, whereas Deutsch in 1985 published a paper [15] explaining the basic idea of quantum Turing machine (QTM) and the universal quantum computer (see Fig. 1.12).

Quantum theory, the Church-Turing principle and the universal quantum computer

DAVID DEUTSCH*

Appeared in *Proceedings of the Royal Society of London A* **400**, pp. 97-117 (1985)[†]

(*Communicated by R. Penrose, F.R.S. — Received 13 July 1984*)

Abstract

It is argued that underlying the Church-Turing hypothesis there is an implicit physical assertion. Here, this assertion is presented explicitly as a physical principle: 'every finitely realizable physical system can be perfectly simulated by a universal model computing machine operating by finite means'. Classical physics and the universal Turing machine, because the former is continuous and the latter discrete, do not obey the principle, at least in the strong form above. A class of model computing machines that is the quantum generalization of the class of Turing machines is described, and it is shown that quantum theory and the 'universal quantum computer' are compatible with the principle. Computing machines resembling the universal quantum computer could, in principle, be built and would have many remarkable properties not reproducible by any Turing machine. These do not include the computation of non-recursive functions, but they do include 'quantum parallelism', a method by which certain probabilistic tasks can be performed faster by a universal quantum computer than by any classical restriction of it. The intuitive explanation of these properties places an intolerable strain on all interpretations of quantum theory other than Everett's. Some of the numerous connections between the quantum theory of computation and the rest of physics are explored. Quantum complexity theory allows a physically more reasonable definition of the 'complexity' or 'knowledge' in a physical system than does classical complexity theory.

*Current address: Centre for Quantum Computation, Clarendon Laboratory, Department of Physics, Parks Road, OX1 3PU Oxford, United Kingdom. Email: david.deutsch@qubit.org
†This version (Summer 1999) was edited and converted to LaTeX by Wim van Dam at the Centre for Quantum Computation. Email: wimvdam@qubit.org

Figure 1.12. David Deutsch and the first page of his 1985 paper

Quantum computers are machines that rely on characteristically quantum phenomena, such as quantum interference and quantum entanglement, in order to perform computation, whereas the classical theory of computation usually refers not to physics but to purely mathematical subjects. A conventional digital computer operates with bits (we may call them *Shannon bits*, since Shannon was the first to use bits to represent information)—the Boolean states 0 and 1—and after each computation step the computer has a definite, exactly measurable state, that is, all bits are in the form 0 or 1 but not both. A quantum computer, a quantum analogue of a digital computer, operates with *quantum bits* (the quantum version of Shannon bit) involving quantum states. The state of a quantum computer is described as a *basis*

vector in a *Hilbert space*,[1] named after the German mathematician David
Hilbert (1862–1943). More formally, we have:

Definition 1.15. A *qubit* is a quantum state $|\Psi\rangle$ of the form

$$|\Psi\rangle = \alpha\,|0\rangle + \beta\,|1\rangle, \tag{1.4}$$

where the amplitudes $\alpha, \beta \in \mathbb{C}$ such that $||\alpha||^2 + ||\beta||^2 = 1$ and $|0\rangle$ and $|1\rangle$
are *basis vectors* of the Hilbert space.

Note that state vectors are written in a special angular bracket notation
called a "ket vector" $|\Psi\rangle$, an expression coined by Paul Dirac who wanted
a shorthand notation for writing formulae that arise in quantum mechanics.
In a quantum computer, each qubit could be represented by the state of a
simple 2-state quantum system such as the spin state of a spin-$\frac{1}{2}$ particle. The
spin of such a particle, when measured, is always found to exist in one of two
possible states $\left|+\frac{1}{2}\right\rangle$ (spin-up) and $\left|-\frac{1}{2}\right\rangle$ (spin-down). This *discreteness* is
called *quantization*. Clearly, the two states can then be used to represent the
binary value 1 and 0 (see Fig. 1.13; by courtesy of Williams and Clearwater
[44]). The main difference between qubits and classical bits is that a bit can

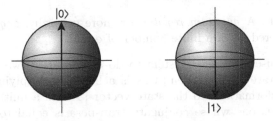

Figure 1.13. A qubit for the binary values 0 and 1

only be set to either 0 or 1, while a qubit $|\Psi\rangle$ can take any (uncountable)
quantum superposition of $|0\rangle$ and $|1\rangle$ (see Fig. 1.14; by courtesy of Williams
and Clearwater [44]). That is, a qubit in a simple 2-state system can have
two states rather than just one allowed at a time as the classical Shannon
bit. Moreover, if a 2-state quantum system can exist in any one of the states
$|0\rangle$ and $|1\rangle$, it can also exist in the *superposed* state

$$|\Psi\rangle = \alpha_1\,|0\rangle + \alpha_2\,|1\rangle. \tag{1.5}$$

---

[1]Hilbert space is defined to be a complete inner-product space. The set of all
sequences $x = (x_1, x_2, \cdots)$ of complex numbers (where $\sum_{i=1}^{\infty} |x_i|^2$ is finite) is a good
example of a Hilbert space, where the sum $x + y$ is defined as $(x_1 + y_1, x_2 + y_2, \cdots)$,
the product ax as $(ax_1, ax_2, \cdots)$, and the inner product as $(x, y) = \sum_{i=1}^{\infty} \overline{x}_i y_i$,
where $\overline{x}_i$ is the complex conjugate of x_i, $x = (x_1, x_2, \cdots)$ and $y = (y_1, y_2, \cdots)$.
In modern quantum mechanics all possible physical states of a system are considered
to correspond to space vectors in a Hilbert space.

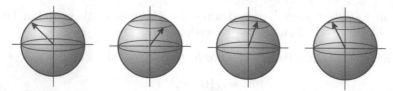

Figure 1.14. Each sphere represents a qubit with the same proportions of the $|0\rangle$ and $|1\rangle$

This is known as the *principle of superposition*. More generally, if a k-state quantum system can exist in any one of the following k eigenstates $|c_1\rangle, |c_1\rangle, \cdots, |c_k\rangle$, it can also exist in the *superposed* state

$$|\Psi\rangle = \sum_{i=0}^{2^k-1} \alpha_i |c_i\rangle, \tag{1.6}$$

where the amplitudes $\alpha_i \in \mathbb{C}$ are such that $\sum_i ||\alpha_i||^2 = 1$ and each $|c_i\rangle$ is a basis vector of the Hilbert space. Once we can encode the binary values 0 and 1 in the states of a physical system, we can make a complete memory of register out of a chain of such systems.

Definition 1.16. A *quantum register*, or more generally, a *quantum computer*, is an ordered set of a finite number of qubits.

In order to use a physical system to do computation, we must be able to change the state of the system; this is achieved by applying a sequence of unitary transformations to the state vector $|\Psi\rangle$ via a unitary matrix (a unitary matrix is one whose conjugate transpose is equal to its inverse). Suppose now a computation is performed on a one-bit quantum computer, then the superposition will be

$$|\Psi\rangle = \alpha |0\rangle + \beta |1\rangle, \tag{1.7}$$

where $\alpha, \beta \in \mathbb{C}$ are such that $||\alpha||^2 + ||\beta||^2 = 1$. The different possible states are $|0\rangle = \begin{pmatrix} 1 \\ 0 \end{pmatrix}$ and $|1\rangle = \begin{pmatrix} 0 \\ 1 \end{pmatrix}$. Let the unitary matrix M be

$$M = \frac{1}{\sqrt{2}} \begin{pmatrix} 1 & 1 \\ -1 & 1 \end{pmatrix}. \tag{1.8}$$

Then the quantum operations on a qubit can be written as follows:

$$M|0\rangle = \frac{1}{\sqrt{2}} \begin{pmatrix} 1 & 1 \\ -1 & 1 \end{pmatrix} \begin{pmatrix} 1 \\ 0 \end{pmatrix} = \frac{1}{\sqrt{2}} |0\rangle - \frac{1}{\sqrt{2}} |1\rangle = |0\rangle$$

$$M|1\rangle = \frac{1}{\sqrt{2}} \begin{pmatrix} 1 & 1 \\ -1 & 1 \end{pmatrix} \begin{pmatrix} 0 \\ 1 \end{pmatrix} = \frac{1}{\sqrt{2}} |0\rangle + \frac{1}{\sqrt{2}} |1\rangle = |1\rangle$$

which is actually the quantum gate (analogous to the classical logic gate):

$$|0\rangle \rightarrow \frac{1}{\sqrt{2}}|0\rangle - \frac{1}{\sqrt{2}}|1\rangle$$

$$|1\rangle \rightarrow \frac{1}{\sqrt{2}}|0\rangle + \frac{1}{\sqrt{2}}|1\rangle.$$

Logic gates can be regarded as logic operators. The NOT operator defined as

$$\text{NOT} = \begin{pmatrix} 0 & 1 \\ 1 & 0 \end{pmatrix}, \tag{1.9}$$

changes the state of its input as follows:

$$\text{NOT}\,|0\rangle = \begin{pmatrix} 0 & 1 \\ 1 & 0 \end{pmatrix} \begin{pmatrix} 1 \\ 0 \end{pmatrix} = \begin{pmatrix} 0 \\ 1 \end{pmatrix} = |1\rangle$$

$$\text{NOT}\,|1\rangle = \begin{pmatrix} 0 & 1 \\ 1 & 0 \end{pmatrix} \begin{pmatrix} 0 \\ 1 \end{pmatrix} = \begin{pmatrix} 1 \\ 0 \end{pmatrix} = |0\rangle.$$

Similarly, we can define the quantum gate of two bits as follows:

$$|00\rangle \rightarrow |00\rangle$$

$$|01\rangle \rightarrow |01\rangle$$

$$|10\rangle \rightarrow \frac{1}{\sqrt{2}}|10\rangle + \frac{1}{\sqrt{2}}|11\rangle$$

$$|11\rangle \rightarrow \frac{1}{\sqrt{2}}|10\rangle - \frac{1}{\sqrt{2}}|11\rangle$$

or equivalently by giving the unitary matrix of the quantum operation:

$$M = \begin{pmatrix} 1 & 0 & 0 & 0 \\ 0 & 1 & 0 & 0 \\ 0 & 0 & \dfrac{1}{\sqrt{2}} & \dfrac{1}{\sqrt{2}} \\ 0 & 0 & \dfrac{1}{\sqrt{2}} & -\dfrac{1}{\sqrt{2}} \end{pmatrix}. \tag{1.10}$$

This matrix is actually the counterpart of the truth table of Boolean logic used for digital computers. Suppose now the computation is in the superposition of the states:

$$\frac{1}{\sqrt{2}}|10\rangle - \frac{1}{\sqrt{2}}|11\rangle$$

or

$$\frac{1}{\sqrt{2}}|10\rangle + \frac{1}{\sqrt{2}}|11\rangle.$$

Then using the unitary transformations defined in (1.10), we have

$$\frac{1}{\sqrt{2}}\,|\,10\rangle - \frac{1}{\sqrt{2}}\,|\,11\rangle = \frac{1}{\sqrt{2}}\left(\frac{1}{\sqrt{2}}\,|\,10\rangle + \frac{1}{\sqrt{2}}\,|\,11\rangle\right)$$

$$-\frac{1}{\sqrt{2}}\left(\frac{1}{\sqrt{2}}\,|\,10\rangle - \frac{1}{\sqrt{2}}\,|\,11\rangle\right)$$

$$= \frac{1}{2}\left(|\,10\rangle + |\,11\rangle\right) - \frac{1}{2}\left(|\,10\rangle - |\,11\rangle\right)$$

$$= |\,11\rangle,$$

$$\frac{1}{\sqrt{2}}\,|\,10\rangle + \frac{1}{\sqrt{2}}\,|\,11\rangle = \frac{1}{2}\left(|\,10\rangle + |\,11\rangle\right) + \frac{1}{2}\left(|\,10\rangle - |\,11\rangle\right)$$

$$= |\,10\rangle.$$

Exercises and Problems for Sect. 1.3

1. Let

$$\mathrm{NOT} = \begin{pmatrix} 0 & 1 \\ 1 & 0 \end{pmatrix}, \quad |\,0\rangle = \begin{pmatrix} 1 \\ 0 \end{pmatrix}.$$

Show that

$$\mathrm{NOT}\,|\,0\rangle = |\,1\rangle.$$

2. Let

$$\mathrm{NOT} = \begin{pmatrix} 0 & 1 \\ 1 & 0 \end{pmatrix}, \quad |\,1\rangle = \begin{pmatrix} 0 \\ 1 \end{pmatrix}.$$

Show that

$$\mathrm{NOT}\,|\,0\rangle = |\,0\rangle.$$

3. Let the action of the $\sqrt{\mathrm{NOT}}$ gate as follows:

$$\sqrt{\mathrm{NOT}} = \begin{pmatrix} \dfrac{1+i}{2} & \dfrac{1-i}{2} \\[2mm] \dfrac{1-i}{2} & -\dfrac{1+i}{2} \end{pmatrix}.$$

Show that

$$\sqrt{\mathrm{NOT}} \cdot \sqrt{\mathrm{NOT}} = \begin{pmatrix} 0 & 1 \\ 1 & 0 \end{pmatrix}.$$

4. Let the conjugate transpose of $\sqrt{\text{NOT}}$, denoted by $(\sqrt{\text{NOT}})^+$, be as follows:

$$(\sqrt{\text{NOT}})^+ = \begin{pmatrix} \dfrac{1-i}{2} & \dfrac{1-i}{2} \\ \dfrac{1+i}{2} & -\dfrac{1-i}{2} \end{pmatrix}.$$

Show that

$$\sqrt{\text{NOT}} \cdot (\sqrt{\text{NOT}})^+ = \begin{pmatrix} 1 & 0 \\ 0 & 1 \end{pmatrix}.$$

5. Let

$$|+\rangle = \frac{1}{\sqrt{2}}(|0\rangle + |1\rangle)$$

$$|-\rangle = \frac{1}{\sqrt{2}}(|0\rangle - |1\rangle)$$

$$|i\rangle = \frac{1}{\sqrt{2}}(|0\rangle + i|1\rangle)$$

$$|-i\rangle = \frac{1}{\sqrt{2}}(|0\rangle - i|1\rangle).$$

Which pairs of expressions for quantum states represent the same state?

(a) $\frac{1}{\sqrt{2}}(|0\rangle + |1\rangle)$ and $\frac{1}{\sqrt{2}}(-|0\rangle + i|1\rangle)$.

(b) $\frac{1}{\sqrt{2}}(|0\rangle + e^{i\pi/4}|1\rangle)$ and $\frac{1}{\sqrt{2}}(e^{-i\pi/4}|0\rangle + |1\rangle)$.

6. Give the set of all values of γ such that following pairs of quantum states are equivalent state:

(a) $|1\rangle$ and $\frac{1}{\sqrt{2}}(|+\rangle + e^{i\gamma}|-\rangle)$.

(b) $\frac{1}{2}|0\rangle - \frac{\sqrt{3}}{2}|1\rangle$ and $e^{i\gamma}\left(\frac{1}{2}|0\rangle - \frac{\sqrt{3}}{2}|1\rangle\right)$.

1.4 Quantum Computability and Complexity

In this section, we shall give a brief introduction to some basic concepts of quantum computability and complexity within the theoretical framework of QTMs.

The first true QTM was proposed in 1985 by Deutsch [15]. A *QTM* is a quantum mechanical generalization of a PTM, in which each cell on the tape can hold a *qubit* (quantum bit) whose state is represented as an arrow contained in a sphere (see Fig. 1.15). Let $\overline{\mathbb{C}}$ be the set consisting of $\alpha \in \mathbb{C}$ such that there is a DTM that computes the real and imaginary parts of α

within 2^{-n} in time polynomial in n, then the QTMs can still be defined as an algebraic system

$$M = (Q, \Sigma, \Gamma, \delta, q_0, \Box, F) \tag{1.11}$$

where

$$\delta : Q \times \Gamma \to \overline{\mathbb{C}}^{Q \times \Gamma \times \{L, R\}}, \tag{1.12}$$

and the rest remains the same as a PTM. Readers are suggested to consult Bernstein and Vazirani [5] for a more detailed discussion of QTMs. QTMs open a new way to model our universe which is quantum physical and offer new features of computation. However, QTMs do not offer more computation power than classical Turing machines. This leads to the following quantitative version of the Church–Turing thesis for quantum computation (see [44]; by courtesy of Williams and Clearwater):

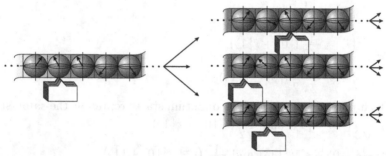

Figure 1.15. A quantum Turing machine

> **The Church–Turing thesis for quantum computation.** Any physical (quantum) computing device can be simulated by a Turing machine in a number of steps polynomial in the resources used by the computing device.

That is, from a computability point of view, a *quantum* Turing machine has no more computation power than a *classical* Turing machine. However, from a computational complexity point of view, a QTM may be more efficient than a classical Turing machine for certain type of computational intractable problems. For example, the integer factorization problem (IFP) and the discrete logarithm problem (DLP) are intractable on classical Turing machines (as everybody knows at present), but they are tractable on QTMs. More precisely, IFP and DLP cannot be solved in polynomial time on a classical computer (classical Turing machine), but can be solved in polynomial time on a quantum computer (QTM).

Remark 1.2. Quantum computers are not just faster versions of classical computers but use a different paradigm for computation. They would speed up the computation of some problems such as IFP and DLP by large factors, but other problems not at all. For quantum computers to be practically useful, we would expect they solve the $\mathcal{NP}$ problems in $\mathcal{P}$. But unfortunately, we do not know this yet. What we know is that quantum computers can solve, e.g., IFP and DLP in $\mathcal{P}$, but IFP and DLP have not been proved in $\mathcal{NP}$.

Just as there are classical complexity classes, so are there quantum complexity classes. As QTMs are generalizations of PTMs, the quantum complexity classes resemble the probabilistic complexity classes. First, we gave the following quantum analogue of classical $\mathcal{P}$:

Definition 1.17. $\mathcal{QP}$ (Quantum Analogue of $\mathcal{P}$) is the class of problems solvable, with certainty, in polynomial time on a QTM.

It can be shown that $\mathcal{P} \subset \mathcal{QP}$ (see Fig. 1.16). That is, the QTM can solve more problems *efficiently* in *worse-case* polynomial time than a classic Turing machine.

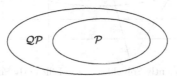

Figure 1.16. Relationship between $\mathcal{QP}$ and $\mathcal{P}$

Similarly, we have the following quantum analogue of classical $\mathcal{ZPP}$.

Definition 1.18. $\mathcal{ZQP}$ (quantum analogue of $\mathcal{ZPP}$) is the class of problems solvable in expected polynomial time with zero-error probability by a QTM.

It is clear that $\mathcal{ZPP} \subset \mathcal{ZQP}$ (see Fig. 1.17).

Figure 1.17. Relationship between $\mathcal{ZQP}$ and $\mathcal{ZPP}$

Definition 1.19. $\mathcal{BQP}$ (quantum analogue of $\mathcal{BPP}$) is the class of problems solvable in polynomial time by a QTM, possibly with a *bounded* probability $\epsilon < 1/3$ of error.

It is known that $\mathcal{P} \subseteq \mathcal{BPP} \subseteq \mathcal{BQP} \subseteq \mathcal{P}$-SPACE, and hence, it is not known whether QTMs are more powerful than PTMs. It is also not known the relationship between $\mathcal{BQP}$ and $\mathcal{NP}$. Figure 1.18 shows the suspected relationship of $\mathcal{BQP}$ to some other well-known classical computational classes.

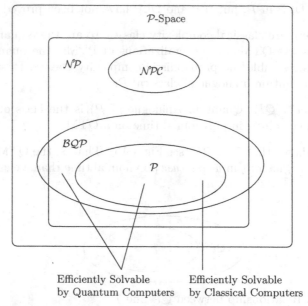

Efficiently Solvable   Efficiently Solvable
by Quantum Computers   by Classical Computers

Figure 1.18. Suspected relationship of $\mathcal{BQP}$ to other classes

Exercises and Problems for Sect. 1.4

1. Explain the complexity classes in the following conjectured containment relations involving classical and quantum computation in Fig. 1.19:

2. Show that
$$\mathcal{P} \subseteq \mathcal{QP} \subseteq \mathcal{BQP}.$$

3. One of the most significant results in quantum computational complexity is that $\mathcal{BQP} \subseteq \mathcal{P}\text{-SPACE}$. Show that
$$\mathcal{BPP} \subseteq \mathcal{BQP} \subseteq \mathcal{P}\text{-SPACE}.$$

4. Show that
$$\mathcal{BQP} \subseteq \mathcal{P}^{\#\mathcal{P}} \subseteq \mathcal{P}\text{-SPACE},$$

where $\mathcal{P}^{\#\mathcal{P}}$ be the set of problems which could be solved in polynomial time if sums of exponentially many terms could be computed efficiently (where these sums must satisfy the requirement that each term is computable in polynomial time).

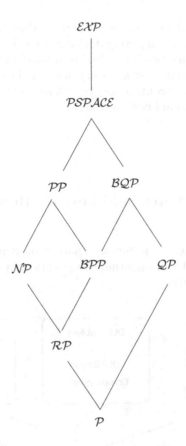

Figure 1.19. Suspected containment relations of complexity classes

5. Show that

$$\mathcal{IP} = \mathcal{P}\text{-SPACE}$$

where $\mathcal{IP}$ is the set of problems having interactive systems and

$$\mathcal{QIP} = \mathcal{P}\text{-SPACE}$$

where $\mathcal{QIP}$ is the set of problems having quantum interactive systems.

6. It is currently not known if a QTM has more computational power than a PTM. Provide evidence (examples of counterexamples) to support the statement that quantum computers do not violate the Church–Turing thesis—any algorithmic process can be simulated by a Turing machine.

7. The Church–Turing thesis (CT), from a computability point of view, can be interpreted as that if a function can be computed by an conceivable hardware system, then it can be computed by a Turing machine. The extended Church–Turing thesis (ECT), from a computational complexity

point of view, makes the stronger assertion that the Turing machine is
also as efficient as any computing device can be. That is, if a function can
be computed by some hardware device in time $T(n)$ for input of size n,
then it can be computed by a Turing machine in time $(T(n))^k$ for fixed
k, depending on the problem. Do you think ECT is valid for quantum
computers and for cloud computation?

1.5 Conclusions, Notes, and Further Reading

This is a preliminary and introductory chapter on computability and complexity of both classical and quantum computation. As the aim of the book
is to study (see Fig. 1.20):

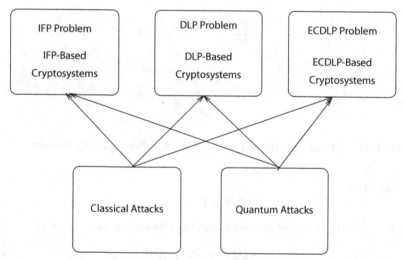

Figure 1.20. Infeasible problems, cryptosystems and the related attacks

1. The three important computationally infeasible (intractably) number-
   theoretic problems, namely, the IFP, the DLP, and the elliptic curve
   discrete logarithm problem (ECDLP), and their related cryptographic
   systems and protocols

2. The classical attacks and quantum attacks on the three infeasible problems and their related cryptographic systems and protocols

We have provided the basic concepts and results for understanding the number-theoretic problems, the cryptographic systems, and the attacks (both classical and quantum) on the problems and the cryptosystems, from a computational point of view.

The theories of computability and computational complexity, including computational Intractability, are intimately connected to cryptography and particularly cryptanalysis. Turing's seminal paper on computable numbers with application to decision problem was published in 1936 [43]; it is in this paper, he proposed the famous Turing machine model. Church's seminal paper on an unsolved problem in elementary number theory was also published in 1936 [7]. So, 1936 is the great year for theoretical computer science. Church also wrote a rather length review paper [8] on Turing paper [43]. The famous Church–Turing thesis was proposed and formulated basically in these three papers. The Cook–Karp thesis was basically proposed and formulated in Cook's 1971 paper [10] and Karp's 1972 paper [25]. These papers, among others, are the founding papers of modern theory of computability and computational complexity. There are a huge number of papers and books devoted to the theory of computability and complexity, including Cook's paper on the $\mathcal{P}$ versus $\mathcal{NP}$ problem [11] and Yao's paper on the Church–Turing thesis and the ECT [51]. The standard references in the field include Hopcroft, Motwani, and Ullman's classical book [24] (now in its 3rd edition) and Garey and Johnson's book on computational intractability [18]. Other excellent and comprehensive books include Lewis and Papadimitrou [28], Linz [29], Papadimitrou [32], and Sipser [41]. More information on number-theoretic computation may be found in [9, 13, 14, 19–21, 26, 34].

Quantum computation is a new paradigm of computation. Quantum computers would speed up some problems by large factors, but not for all problems. In fact, as far as we know at present, quantum computation does not violate the Church–Turing thesis, and quantum computers do not offer more computational power than classical computers . The first person to systematically study quantum computation is possibly the 1965 Nobel Laureate Richard Feynman (see Feynman [16, 17]). The following references provide more information on quantum computing, including quantum computability and quantum complexity: [2, 4, 6, 22, 23, 27, 30, 31, 33, 35–39, 42, 45–50].

There is a special section on quantum computation in SIAM Journal, Volume 26, Number 5, October 1997, with some of the classic papers in the field by Bernstein and Vazirani [5] on quantum complexity theory, Simon [40] on the power of quantum computation, Shor [36] on polynomial-time quantum algorithms for IFP and DLP, Bennett [3] on strengths and weaknesses of quantum computing, and Adleman et al. [1] on quantum computability.

REFERENCES

[1]   L.M. Adleman, J. DeMarrais, M.-D.A. Huang, Quantum computability. SIAM
      J. Comput. **26**(5), 1524–1540 (1996)

[2]   P. Benioff, The computer as a physical system – a microscopic quantum me-
      chanical Hamiltonian model of computers as represented by turing machines.
      J. Stat. Phys. **22**, 563–591 (1980)

[3]   C.H. Bennett, Strengths and weakness of quantum computing. SIAM J. Com-
      put. **26**(5), 1510–1523 (1997)

[4]   C.H. Bennett, D.P. DiVincenzo, Quantum information and computation. Na-
      ture **404**, 247–255 (2000)

[5]   E. Bernstein, U. Vazirani, Quantum complexity theory. SIAM J. Comput.
      26(5), 1411–1473 (1997)

[6]   I.L. Change, R. Laflamme, P. Shor, W.H. Zurek, Quantum computers, factor-
      ing, and decoherence. Science **270**, 1633–1635 (1995)

[7]   A. Church, An unsolved problem of elementary number theory. Am. J. Math.
      58, 345–363 (1936)

[8]   A. Church, Book review: on computable numbers, with an application to the
      Entscheidungsproblem by Turing. J. Symbolic Log. **2**, 42–43 (1937)

[9]   H. Cohen, in *A Course in Computational Algebraic Number Theory*. Graduate
      Texts in Mathematics, vol. 138 (Springer, Berlin, 1993)

[10] S. Cook, The complexity of theorem-proving procedures, in *Proceedings of the
      3rd Annual ACM Symposium on the Theory of Computing*, New York, 1971,
      pp. 151–158

[11] S. Cook, The importance of the P versus NP question. J. ACM **50**(1), 27–29
      (2003)

[12] S. Cook, The P versus NP problem, in *The Millennium Prize Problems*, ed. by
      J. Carlson, A. Jaffe, A. Wiles. (Clay Mathematics Institute/American Math-
      ematical Society, Providence, 2006), pp. 87–104

[13] T.H. Cormen, C.E. Ceiserson, R.L. Rivest, *Introduction to Algorithms*, 3rd
      edn. (MIT, Cambridge, 2009)

[14] R. Crandall, C. Pomerance, *Prime Numbers – A Computational Perspective*,
      2nd edn. (Springer, Berlin, 2005)

[15] D. Deutsch, Quantum theory, the Church–Turing principle and the universal
      quantum computer. Proc. R. Soc. Lond. Ser. A **400**, 96–117 (1985)

[16] R.P. Feynman, Simulating physics with computers. Int. J. Theor. Phys. **21**,
      467–488 (1982)

[17] R.P. Feynman, in *Feynman Lectures on Computation*, ed. by A.J.G. Hey, R.W.
      Allen (Addison-Wesley, Reading, 1996)

[18] M.R. Garey, D.S. Johnson, *Computers and Intractability – A Guide to the
      Theory of NP-Completeness* (W.H. Freeman and Company, San Francisco,
      1979)

[19] O. Goldreich, *Foundations of Cryptography: Basic Tools* (Cambridge Univer-
      sity Press, Cambridge, 2001)

[20] O. Goldreich, *Foundations of Cryptography: Basic Applications* (Cambridge
      University Press, Cambridge, 2004)

[21] O. Goldreich, *P, NP, and NP-Completeness* (Cambridge University Press, Cambridge, 2010)

[22] J. Grustka, *Quantum Computing* (McGraw-Hill, New York, 1999)

[23] M. Hirvensalo, *Quantum Computing*, 2nd edn. (Springer, Berlin, 2004)

[24] J. Hopcroft, R. Motwani, J. Ullman, *Introduction to Automata Theory, Languages, and Computation*, 3rd edn. (Addison-Wesley, Reading, 2007)

[25] R. Karp, Reducibility among combinatorial problems, in *Complexity of Computer Computations*, ed. by R.E. Miller, J.W. Thatcher (Plenum, New York, 1972), pp. 85–103

[26] D.E. Knuth, *The Art of Computer Programming II – Seminumerical Algorithms*, 3rd edn. (Addison-Wesley, Reading, 1998)

[27] M. Le Bellac, *A Short Introduction to Quantum Information and Quantum Computation* (Cambridge University Press, Cambridge, 2005)

[28] H.R. Lewis, C.H. Papadimitrou, *Elements of the Theory of Computation* (Prentice-Hall, Englewood Cliffs, 1998)

[29] P. Linz, *An Introduction to Formal Languages and Automata*, 5th edn. (Jones and Bartlett Publishers, Burlington, Massachusetts, 2011)

[30] N.D. Mermin, *Quantum Computer Science* (Cambridge University Press, Cambridge, 2007)

[31] M.A. Nielson, I.L. Chuang, *Quantum Computation and Quantum Information*, 10th Anniversary edn. (Cambridge University Press, Cambridge, 2010)

[32] C.H. Papadimitrou, *Computational Complexity* (Addison Wesley, Reading, 1994)

[33] E. Rieffel, W. Polak, *Quantum Computing: A Gentle Introduction* (MIT, Cambridge, 2011)

[34] H. Riesel, *Prime Numbers and Computer Methods for Factorization* (Birkhäuser, Boston, 1990)

[35] P. Shor, Algorithms for quantum computation: discrete logarithms and factoring, in *Proceedings of 35th Annual Symposium on Foundations of Computer Science* (IEEE Computer Society, Silver Spring, 1994), pp. 124–134

[36] P. Shor, Polynomial-Tme algorithms for prime factorization and discrete logarithms on a quantum computer. SIAM J. Comput. **26**(5), 1411–1473 (1997)

[37] P. Shor, Quantum computing. Documenta Math. Extra Volume ICM **I**, 467–486 (1998)

[38] P. Shor, Introduction to quantum algorithms. AMS Proc. Symp. Appl. Math. **58**, 17 (2002)

[39] P. Shor, Why haven't more quantum algorithms been found? J. ACM **50**(1), 87–90 (2003)

[40] D.R. Simon, On the power of quantum computation. SIAM J. Comput. **26**(5), 1474–1483 (1997)

[41] M. Sipser, *Introduction to the Theory of Computation*, 2nd edn. (Thomson, Boston, 2006)

[42] W. Trappe, L. Washington, *Introduction to Cryptography with Coding Theory*, 2nd edn. (Prentice-Hall, Englewood Cliffs, 2006)

[43] A. Turing, On computable numbers, with an application to the Entscheidungsproblem. Proc. Lond. Math. Soc. Ser. 2 **42**, 230–260 (1937); **43**, 544–546 (1937)

[44] C.P. Williams, S.H. Clearwater, in *Explorations in Quantum Computation.* The Electronic Library of Science (TELOS) (Springer, Berlin, 1998)

[45] U.V. Vazirani, On the power of quantum computation. Phil. Trans. R. Soc. Lond. **A356**, 1759–1768 (1998)

[46] U.V. Vazirani, Fourier transforms and quantum computation, in *Proceedings of Theoretical Aspects of Computer Science* (Springer, Berlin, 2000), pp. 208–220

[47] U.V. Vazirani, A survey of quantum complexity theory. AMS Proc. Symp. Appl. Math. **58**, 28 (2002)

[48] J. Watrous, Quantum computational complexity, in *Encyclopedia of Complexity and System Science* (Springer, Berlin, 2009), pp. 7174–7201

[49] C.P. Williams, *Explorations in Quantum Computation*, 2nd edn. (Springer, Berlin, 2011)

[50] N.S. Yanofsky, M.A. Mannucci, *Quantum Computing for Computer Scientists* (Cambridge University Press, Cambridge, 2008)

[51] A. Yao, Classical physics and the Church Turing thesis. J. ACM **50**(1), 100–105 (2003)

2. Quantum Attacks on IFP-Based Cryptosystems

If you don't work on important problems, it's not likely that you'll do important work.

RICHARD HAMMING (1915–1998)
The 1968 Turing Award Recipient

In this chapter we shall first study the integer factorization problem (IFP) and the classical solutions to IFP, then we shall discuss the IFP-based cryptography whose security relies on the infeasibility of the IFP problem, and finally, we shall introduce some quantum algorithms for attacking both IFP and IFP-based cryptography.

2.1 IFP and Classical Solutions to IFP

Fundamental Theorem of Arithmetic

In mathematics, there are many fundamental theorems such as fundamental theorem of geometry, fundamental theorem of algebra, and fundamental theorem of calculus. The fundamental theorem of arithmetic (FTA) may be regarded as the first and most important fundamental theorem in mathematics, stating as follows.

Theorem 2.1 (FTA). Any positive integer $n > 1$ can be written uniquely as the following standard prime factorization form:

$$n = p_1^{\alpha_1} p_2^{\alpha_2} \cdots p_k^{\alpha_k}, \qquad (2.1)$$

where $p_1 < p_2 < \cdots < p_k$ are primes and $\alpha_1, \alpha_2, \cdots, \alpha_k$ are positive integers.

S.Y. Yan, *Quantum Attacks on Public-Key Cryptosystems*,
DOI 10.1007/978-1-4419-7722-9_2,
© Springer Science+Business Media, LLC 2013

Integer Factorization Problem

The idea of FTA can be traced to Euclid's *Elements* [25], but it was first clearly stated and proved by Gauss [29] in his *Disquisitiones*. According to FTA, any positive integer can be uniquely written as its prime decomposition form, say, for example,

$$12345678987654321 = 3^4 \cdot 37^2 \cdot 333667^2.$$

So, we can define the prime factorization problem (PFP) as follows:

$$\text{PFP} \overset{\text{def}}{=} \begin{cases} \text{Input}: & n \in \mathbb{Z}_{>1} \text{ and } n \notin \text{Primes} \\ \text{Output}: & n = p_1^{\alpha_1} p_2^{\alpha_2} \cdots p_k^{\alpha_k} \end{cases} \tag{2.2}$$

The solution to PFP is actually involved in the solutions of two other problems: the primality testing problem (PTP) and the IFP, which can be described as follows:

$$\text{PTP} \overset{\text{def}}{=} \begin{cases} \text{Input}: & n \in \mathbb{Z}_{>1} \\ \text{Output}: & \begin{cases} \text{Yes}, & n \in \text{Primes} \\ \text{No}, & \text{Otherwise} \end{cases} \end{cases} \tag{2.3}$$

and

$$\text{IFP} \overset{\text{def}}{=} \begin{cases} \text{Input}: & n \in \mathbb{Z}_{>1} \text{ and } n \notin \text{Primes} \\ \text{Output}: & 1 < f < n \ (f \text{ is a nontrivial factor of } n). \end{cases} \tag{2.4}$$

So, to solve PFP, one just needs to recursively execute the following two algorithms:

1. Algorithm for PTP

2. Algorithm for IFP

That is,

$$\text{PFP} \overset{\text{def}}{=} \overset{\curvearrowright}{\text{PTP}} \oplus \overset{\curvearrowright}{\text{IFP}}.$$

For example, if we wish to factor the integer 123457913315, the recursive process may be shown in Fig. 2.1. Since PTP can be solved easily in polynomial time [3], we shall only concentrate on the solutions to IFP.

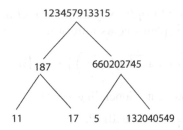

Figure 2.1. Prime factorization of 123457913315

Methods for Integer Factorization

There are many methods and algorithms for factoring a large integer. If we are concerned with the determinism of the algorithms, then there are two types of factoring algorithms:

1. Deterministic factoring algorithms
2. Probabilistic factoring algorithms

However, if we are more concerned with the form and the property of the integers to be factored, then there are two types factoring methods or algorithms:

1. General-purpose factoring algorithms: The running time depends mainly on the size of N, the number to be factored, and is not strongly dependent on the size of the factor p found. Examples are:

   (a) *Lehman's method* [44], which has a rigorous worst-case running time bound $\mathcal{O}\left(n^{1/3+\epsilon}\right)$.

   (b) *Euler's factoring method* [49], which has deterministic running time $\mathcal{O}\left(n^{1/3+\epsilon}\right)$.

   (c) *Shanks' SQUare FOrm Factorization method [68]* SQUFOF, which has expected running time $\mathcal{O}\left(n^{1/4}\right)$.

   (d) *The FFT-based factoring methods of Pollard and Strassen* [58, 75] which have deterministic running time $\mathcal{O}\left(n^{1/4+\epsilon}\right)$.

   (e) *The lattice-based factoring methods of Coppersmith* [18], which has deterministic running time $\mathcal{O}\left(n^{1/4+\epsilon}\right)$.

   (f) *Shanks' class group method* [67], which has running time $\mathcal{O}\left(n^{1/5+\epsilon}\right)$, assuming the extended Riemann's hypothesis (ERH).

   (g) *Continued FRACtion (CFRAC) method* [55], which under plausible assumptions has expected running time

   $$\mathcal{O}\left(\exp\left(c\sqrt{\log n \log\log n}\,\right)\right) = \mathcal{O}\left(n^{c\sqrt{\log\log n/\log n}}\,\right),$$

   where c is a constant (depending on the details of the algorithm); usually $c = \sqrt{2} \approx 1.414213562$.

(h) *Quadratic sieve/Multiple polynomial quadratic sieve (QS/MPQS)*
    [60], which under plausible assumptions has expected running time

$$\mathcal{O}\left(\exp\left(c\sqrt{\log n \log\log n}\,\right)\right) = \mathcal{O}\left(n^{c\sqrt{\log\log n/\log n}}\,\right),$$

where c is a constant (depending on the details of the algorithm);
usually $c = \dfrac{3}{2\sqrt{2}} \approx 1.060660172$.

(i) *Number field sieve (NFS)* [46], which under plausible assumptions
    has the expected running time

$$\mathcal{O}\left(\exp\left(c\sqrt[3]{\log n}\sqrt[3]{(\log\log n)^2}\,\right)\right),$$

where $c = (64/9)^{1/3} \approx 1.922999427$ if GNFS (a general version of
NFS) is used to factor an arbitrary integer n, whereas $c = (32/9)^{1/3} \approx$
1.526285657 if SNFS (a special version of NFS) is used to factor a
special integer n such as $n = r^e \pm s$, where r and s are small, $r > 1$,
and e is large. This is substantially and asymptotically faster than
any other currently known factoring method.

2. Special purpose factoring algorithms: The running time depends mainly
   on the size of p (the factor found) of n. (We can assume that $p \leqslant \sqrt{n}$.)
   Examples are:

(a) *Trial division* [41], which has running time $\mathcal{O}\left(p(\log n)^2\right)$.

(b) *Pollard's ρ-method* [10, 59] (also known as Pollard's "*rho*" algo-
    rithm), which under plausible assumptions has expected running time
    $\mathcal{O}\left(p^{1/2}(\log n)^2\right)$.

(c) *Pollard's $p-1$ method* [58], which runs in $\mathcal{O}(B\log B(\log n)^2)$, where
    B is the smooth bound; larger values of B make it run more slowly,
    but are more likely to produce a factor of n.

(d) *Lenstra's elliptic curve method (ECM)* [47], which under plausible
    assumptions has expected running time

$$\mathcal{O}\left(\exp\left(c\sqrt{\log p \log\log p}\,\right)\cdot(\log n)^2\right),$$

where $c \approx 2$ is a constant (depending on the details of the algorithm).
The term $\mathcal{O}\left((\log n)^2\right)$ is for the cost of performing arithmetic operations
on numbers which are $\mathcal{O}(\log n)$ or $\mathcal{O}\left((\log n)^2\right)$ bits long; the second can
be theoretically replaced by $\mathcal{O}\left((\log n)^{1+\epsilon}\right)$ for any $\epsilon > 0$.

Note that there is a quantum factoring algorithm, first proposed by Shor
[70], which can run in polynomial time

$$\mathcal{O}((\log n)^{2+\epsilon}).$$

However, this quantum algorithm requires to be run on a quantum computer, which is not available at present.

In practice, algorithms in both categories are important. It is sometimes very difficult to say whether one method is better than another, but it is generally worth attempting to find small factors with algorithms in the second class before using the algorithms in the first class. That is, we could first try the *trial division algorithm*, then use some other method such as NFS. This fact shows that the trial division method is still useful for integer factorization, even though it is simple. In this chapter we shall introduce some most the useful and widely used factoring algorithms.

From a computational complexity point of view, the IFP is an infeasible (intractable) problem, since there is no polynomial-time algorithm for solving it; all the existing algorithms for IFP run in subexponential-time or above (see Fig. 2.2). Note that there is a quantum algorithm proposed by Shor [70]

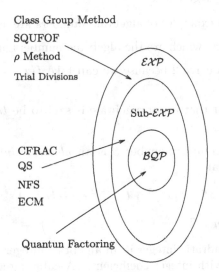

Figure 2.2. Algorithms/methods for IFP

for IFP that can be run in polynomial time, but it needs to be run on a practical quantum computer which does not exist at present.

NFS Factoring

A fundamental idea of many modern general-purpose algorithms for factoring n is to find a suitable pair if (x, y) such that

$$x^2 \equiv y^2 \ (\text{mod} \ n) \ \text{ but } \ x \not\equiv \pm y \ (\text{mod} \ n)$$

then there is a good chance to factor n:

$$\text{Prob}(\gcd(x \pm y, n) = (f_1, f_2), \ 1 < f_1, f_2 < n) > \frac{1}{2}.$$

In practice, the asymptotically fastest general-purpose factoring algorithm is the NFS, and it runs in expected subexponential-time:

$$\mathcal{O}(\exp(c(\log n)^{1/3}(\log \log n)^{2/3})).$$

Definition 2.1. A complex number α is an *algebraic number* if it is a root of a polynomial

$$f(x) = a_0 x^k + a_1 x^{k-1} + a_2 x^{k-2} \cdots + a_k = 0 \qquad (2.5)$$

where $a_0, a_1, a_2, \ldots, a_k \in \mathbb{Q}$ and $a_0 \neq 0$. If $f(x)$ is irreducible over $\mathbb{Q}$ and $a_0 \neq 0$, then k is the degree of x.

Example 2.1. Two examples of algebraic numbers are as follows:

1 Rational numbers, which are the algebraic numbers of degree 1

2 $\sqrt{2}$, which is of degree 2 because we can take $f(x) = x^2 - 2 = 0$ ($\sqrt{2}$ is irrational)

Any complex number that is not algebraic is said to be *transcendental* such as π and e.

Definition 2.2. A complex number β is an *algebraic integer* if it is a root of a monic polynomial

$$x^k + b_1 x^{k-1} + b_2 x^{k-2} \cdots + b_k = 0 \qquad (2.6)$$

where $b_0, b_1, b_2, \ldots, b_k \in \mathbb{Z}$.

Remark 2.1. A quadratic integer is an algebraic integer satisfying a monic quadratic equation with integer coefficients. A cubic integer is an algebraic integer satisfying a monic cubic equation with integer coefficients.

Example 2.2. Some examples of algebraic integers are as follows:

1 Ordinary (rational) integers, which are the algebraic integers of degree 1. That is, they satisfy the monic equations $x - a = 0$ for $a \in \mathbb{Z}$.

2 $\sqrt[3]{2}$ and $\sqrt[5]{3}$, because they satisfy the monic equations $x^3 - 2 = 0$ and $x^3 - 5 = 0$, respectively.

3 $(-1 + \sqrt{-3})/2$, because it satisfies $x^2 + x + 1 = 0$.

4 Gaussian integer $a + b\sqrt{-1}$, with $a, b \in \mathbb{Z}$.

Clearly, every algebraic integer is an algebraic number, but the converse is not true.

Proposition 2.1. A rational number $r \in \mathbb{Q}$ is an algebraic integer if and only if $r \in \mathbb{Z}$.

Proof. If $r \in \mathbb{Z}$, then r is a root of $x - r = 0$. Thus, r is an algebraic integer. Now, suppose that $r \in \mathbb{Q}$ and r is an algebraic integer (i.e., $r = c/d$ is a root of (2.6), where $c, d \in \mathbb{Z}$; we may assume $\gcd(c, d) = 1$). Substituting c/d into (2.6) and multiplying both sides by d^n, we get

$$c^k + b_1 c^{k-1} d + b_2 c^{k-2} d^2 \cdots + b_k d^k = 0.$$

It follows that $d \mid c^k$ and $d \mid c$ (since $\gcd(c, d) = 1$). Again, since $\gcd(c, d) = 1$, it follows that $d = \pm 1$. Hence, $r = c/d \in \mathbb{Z}$. It follows, for example, that $2/5$ is an algebraic number but not an algebraic integer. □

Remark 2.2. The elements of $\mathbb{Z}$ are the only rational numbers that are algebraic integers. We shall refer to the elements of $\mathbb{Z}$ as *rational integers* when we need to distinguish them from other algebraic integers that are not rational. For example, $\sqrt{2}$ is an algebraic integer but not a rational integer.

The most interesting results concerned with the algebraic numbers and algebraic integers are the following theorem.

Theorem 2.2. The set of algebraic numbers forms a field, and the set of algebraic integers forms a ring.

Proof. See pp 67–68 of Ireland and Rosen [38]. □

Lemma 2.1. Let $f(x)$ be an irreducible monic polynomial of degree d over integers and m an integer such that $f(m) \equiv 0 \pmod{n}$. Let α be a complex root of $f(x)$ and $\mathbb{Z}[\alpha]$ the set of all polynomials in α with integer coefficients. Then there exists a unique mapping $\Phi : \mathbb{Z}[\alpha] \mapsto \mathbb{Z}_n$ satisfying:

1 $\Phi(ab) = \Phi(a)\Phi(b), \quad \forall a, b \in \mathbb{Z}[\alpha]$.

2 $\Phi(a + b) = \Phi(a) + \Phi(b), \quad \forall a, b \in \mathbb{Z}[\alpha]$.

3 $\Phi(za) = z\Phi(a), \quad \forall a \in \mathbb{Z}[\alpha], z \in \mathbb{Z}$.

4 $\Phi(1) = 1$.

5 $\Phi(\alpha) = m \pmod{n}$.

Now, we are in a position to introduce the NFS. Note that there are two main types of NFS: NFS (general NFS) for general numbers and SNFS (special NFS) for numbers with special forms. The idea, however, behind the GNFS and SNFS is the same:

1. Find a monic irreducible polynomial $f(x)$ of degree d in $\mathbb{Z}[x]$ and an integer m such that $f(m) \equiv 0 \pmod{n}$.

2. Let $\alpha \in \mathbb{C}$ be an algebraic number that is the root of $f(x)$, and denote the set of polynomials in α with integer coefficients as $\mathbb{Z}[\alpha]$.

3. Define the mapping (ring homomorphism): $\Phi : \mathbb{Z}[\alpha] \mapsto \mathbb{Z}_n$ via $\Phi(\alpha) = m$ which ensures that for any $f(x) \in \mathbb{Z}[x]$, we have $\Phi(f(\alpha)) \equiv f(m) \pmod{n}$.

4. Find a finite set U of coprime integers (a, b) such that

$$\prod_{(a,b)\in U} (a - b\alpha) = \beta^2, \quad \prod_{(a,b)\in U} (a - bm) = y^2$$

for $\beta \in \mathbb{Z}[\alpha]$ and $y \in \mathbb{Z}$. Let $x = \Phi(\beta)$. Then

$$x^2 \equiv \Phi(\beta)\Phi(\beta)$$

$$\equiv \Phi(\beta^2)$$

$$\equiv \Phi\left(\prod_{(a,b)\in U} (a - b\alpha)\right)$$

$$\equiv \prod_{(a,b)\in U} \Phi(a - b\alpha)$$

$$\equiv \prod_{(a,b)\in U} (a - bm)$$

$$\equiv y^2 \pmod{n}$$

which is of the required form of the factoring congruence, and hopefully, a factor of n can be found by calculating $\gcd(x \pm y, n)$.

There are many ways to implement the above idea, all of which follow the same pattern as we discussed previously in CFRAC and QS/MPQS: By a sieving process, one first tries to find congruences modulo n by working over a factor base, and then do a Gaussian elimination over $\mathbb{Z}/2\mathbb{Z}$ to obtain a congruence of squares $x^2 \equiv y^2 \pmod{n}$. We give in the following a brief description of the NFS algorithm [54].

Algorithm 2.1. Given an odd positive integer n, the NFS algorithm has the following four main steps in factoring n:

[1] (Polynomials Selection) Select two irreducible polynomials $f(x)$ and $g(x)$ with small integer coefficients for which there exists an integer m such that

$$f(m) \equiv g(m) \equiv 0 \pmod{n} \tag{2.7}$$

The polynomials should not have a common factor over $\mathbb{Q}$.

[2] (Sieving) Let α be a complex root of f and β a complex root of g. Find pairs (a, b) with $\gcd(a, b) = 1$ such that the integral norms of $a - b\alpha$ and $a - b\beta$

$$N(a - b\alpha) = b^{\deg(f)} f(a/b), \qquad N(a - b\beta) = b^{\deg(g)} g(a/b) \qquad (2.8)$$

are smooth with respect to a chosen factor base. (The principal ideals $a - b\alpha$ and $a - b\beta$ factor into products of prime ideals in the number field $\mathbb{Q}(\alpha)$ and $\mathbb{Q}(\beta)$, respectively.)

[3] (Linear Algebra) Use techniques of linear algebra to find a set $U = \{a_i, b_i\}$ of indices such that the two products

$$\prod_U (a_i - b_i\alpha), \qquad \prod_U (a_i - b_i\beta) \qquad (2.9)$$

are both squares of products of prime ideals.

[4] (Square root) Use the set S in (2.9) to find an algebraic numbers $\alpha' \in \mathbb{Q}(\alpha)$ and $\beta' \in \mathbb{Q}(\beta)$ such that

$$(\alpha')^2 = \prod_U (a_i - b_i\alpha), \qquad (\beta')^2 = \prod_U (a_i - b_i\beta) \qquad (2.10)$$

Define $\Phi_\alpha : \mathbb{Q}(\alpha) \to \mathbb{Z}_n$ and $\Phi_\beta : \mathbb{Q}(\beta) \to \mathbb{Z}_n$ via $\Phi_\alpha(\alpha) = \Phi_\beta(\beta) = m$, where m is the common root of both f and g. Then

$$
\begin{aligned}
x^2 &\equiv \Phi_\alpha(\alpha')\Phi_\alpha(\alpha') \\
&\equiv \Phi_\alpha((\alpha')^2) \\
&\equiv \Phi_\alpha\left(\prod_{i \in U}(a_i - b_i\alpha)\right) \\
&\equiv \prod_U \Phi_\alpha(a_i - b_i\alpha) \\
&\equiv \prod_U (a_i - b_i m) \\
&\equiv \Phi_\beta(\beta')^2 \\
&\equiv y^2 \qquad (\bmod\ n)
\end{aligned}
$$

which is of the required form of the factoring congruence, and hopefully, a factor of N can be found by calculating $\gcd(x \pm y, n)$.

Example 2.3. We first give a rather simple NFS factoring example. Let $n = 14885 = 5 \cdot 13 \cdot 229 = 122^2 + 1$. So we put $f(x) = x^2 + 1$ and $m = 122$, such that

$$f(x) \equiv f(m) \equiv 0 \pmod{n}.$$

If we choose $|a|, |b| \leqslant 50$, then we can easily find (by sieving) that

(a, b)	$\mathrm{Norm}(a + bi)$	$a + bm$
$\vdots$	$\vdots$	$\vdots$
$(-49, 49)$	$4802 = 2 \cdot 7^4$	$5929 = 7^2 \cdot 11^2$
$\vdots$	$\vdots$	$\vdots$
$(-41, 1)$	$1682 = 2 \cdot 29^2$	$81 = 3^4$
$\vdots$	$\vdots$	$\vdots$

(Readers should be able to find many such pairs of (a_i, b_i) in the interval that are smooth up to, e.g., 29.) So, we have

$$
\begin{aligned}
(49 + 49i)(-41 + i) &= (49 - 21i)^2, \\
f(49 - 21i) &= 49 - 21m \\
&= 49 - 21 \cdot 122 \\
&= -2513 \to x, \\
5929 \cdot 81 &= (2^2 \cdot 7 \cdot 11)^2 \\
&= 693^2 \\
&\to y = 693.
\end{aligned}
$$

Thus,

$$
\begin{aligned}
\gcd(x \pm y, n) &= \gcd(-2513 \pm 693, 14885) \\
&= (65, 229).
\end{aligned}
$$

In the same way, if we wish to factor $n = 84101 = 290^2 + 1$, then we let $m = 290$ and $f(x) = x^2 + 1$ so that

$$f(x) \equiv f(m) \equiv 0 \pmod{n}.$$

We tabulate the sieving process as follows:

(a, b)	Norm$(a + bi)$	$a + bm$
$\vdots$	$\vdots$	$\vdots$
$-50, 1$	$2501 = 41 \cdot 61$	$240 = 2^4 \cdot 3 \cdot 5$
$\vdots$	$\vdots$	$\vdots$
$-50, 3$	$2509 = 13 \cdot 193$	$820 = 2^2 \cdot 5 \cdot 41$
$\vdots$	$\vdots$	$\vdots$
$-49, 43$	$4250 = 2 \cdot 5^3 \cdot 17$	$12421 = 12421$
$\vdots$	$\vdots$	$\vdots$
$-38, 1$	$1445 = 5 \cdot 17^2$	$252 = 2^2 \cdot 3^2 \cdot 7$
$\vdots$	$\vdots$	$\vdots$
$-22, 19$	$845 = 5 \cdot 13^2$	$5488 = 2^4 \cdot 7^3$
$\vdots$	$\vdots$	$\vdots$
$-118, 11$	$14045 = 5 \cdot 53^2$	$3072 = 2^{10} \cdot 3$
$\vdots$	$\vdots$	$\vdots$
$218, 59$	$51005 = 5 \cdot 101^2$	$17328 = 2^4 \cdot 3 \cdot 19^2$
$\vdots$	$\vdots$	$\vdots$

Clearly, $-38 + i$ and $-22 + 19i$ can produce a product square, since

$$
\begin{aligned}
(-38 + i)(-22 + 19i) &= (31 - 12i)^2, \\
f(31 - 12i) &= 31 - 12m \\
&= -3449 \to x, \\
252 \cdot 5488 &= (2^3 \cdot 3 \cdot 7^2)^2 \\
&= 1176^2, \\
&\to y = 1176, \\
\gcd(x \pm y, n) &= \gcd(-3449 \pm 1176, 84101) \\
&= (2273, 37).
\end{aligned}
$$

In fact, $84101 = 2273 \times 37$. Note that $-118 + 11i$ and $218 + 59i$ can also produce a product square, since

$$
\begin{aligned}
(-118 + 11i)(218 + 59i) &= (14 - 163i)^2, \\
f(14 - 163i) &= 14 - 163m \\
&= -47256 \rightarrow x, \\
3071 \cdot 173288 &= (2^7 \cdot 3 \cdot 19)^2 \\
&= 7296^2, \\
&\rightarrow y = 7296, \\
\gcd(x \pm y, n) &= \gcd(-47256 \pm 7296, 84101) \\
&= (37, 2273).
\end{aligned}
$$

Example 2.4. Next, we present a little bit more complicated example. Use NFS to factor $n = 1098413$. First, notice that $n = 1098413 = 12 \cdot 45^3 + 17^3$, which is in a special form and can be factored by using SNFS.

[1] (Polynomials Selection) Select the two irreducible polynomials $f(x)$ and $g(x)$ and the integer m as follows:

$$
m = \frac{17}{45},
$$

$$
f(x) = x^3 + 12 \implies f(m) = \left(\frac{17}{45}\right)^3 + 12 \equiv 0 \pmod{n},
$$

$$
g(x) = 45x - 17 \implies g(m) = 45 \left(\frac{17}{45}\right) - 17 \equiv 0 \pmod{n}.
$$

[2] (Sieving) Suppose after sieving, we get $U = \{a_i, b_i\}$ as follows:

$$
U = \{(6, -1), (3, 2), (-7, 3), (1, 3), (-2, 5), (-3, 8), (9, 10)\}.
$$

That is, the chosen polynomial that produces a product square can be constructed as follows (as an exercise, readers may wish to choose some other polynomial which can also produce a product square):

$$
\prod_U (a_i + b_i x) = (6-x)(3+2x)(-7+3x)(1+3x)(-2+5x)(-3+8x)(9+10x).
$$

Let $\alpha = \sqrt[3]{-12}$ and $\beta = \frac{17}{45}$. Then

$$\prod_U (a - b\alpha) \quad = \quad 7400772 + 1138236\alpha - 10549\alpha^2$$

$$= \quad (2694 + 213\alpha - 28\alpha^2)^2$$

$$= \quad \left(\frac{5610203}{2025}\right)$$

$$= \quad 270729^2,$$

$$\prod_U (a - b\beta) \quad = \quad \frac{2^8 \cdot 11^2 \cdot 13^2 \cdot 23^2}{3^{12} \cdot 5^4}$$

$$= \quad \left(\frac{52624}{18225}\right)^2$$

$$= \quad 875539^2.$$

So, we get the required square of congruence:

$$270729^2 \equiv 875539^2 \ (\text{mod } 1098413).$$

Thus,

$$\gcd(270729 \pm 875539, 1098413) = (563, 1951).$$

That is,

$$1098413 = 563 \cdot 1951.$$

Example 2.5. We give some large factoring examples using NFS.

1 SNFS examples: One of the largest numbers factored by SNFS is

$$n = (12^{167} + 1)/13 = p_{75} \times p_{105}$$

It was announced by P. Montgomery, S. Cavallar, and H. te Riele at CWI in Amsterdam on 3 September 1997. They used the polynomials $f(x) = x^5 - 144$ and $g(x) = 12^{33}x + 1$ with common root $m \equiv 12^{134} \ (\text{mod } n)$. The factor base bound was 4.8 million for f and 12 million for g. Both large prime bounds were 150 million, with two large primes allowed on each side. They sieved over $|a| \leqslant 8.4$ million and $0 < b \leqslant 2.5$ million. The sieving lasted 10.3 calendar days; 85 SGI machines at CWI contributed a combined 13027719 relations in 560 machine-days. It took 1.6 more calendar days to process the data. This processing included 16 CPU-hours on a Cray C90 at SARA in Amsterdam to process a 1969262×1986500 matrix with 57942503 nonzero entries. The other large number factorized by using SNFS is the 9th Fermat number:

$$F_9 = 2^{2^9} + 1 = 2^{512} + 1 = 2424833 \cdot p_{49} \cdot p_{99},$$

a number with 155 digits; it was completely factored in April 1990. The most wanted factoring number of special form at present is the 12th Fermat number $F_{12} = 2^{2^{12}} + 1$; we only know its partial prime factorization:

$$F_{12} = 114689 \cdot 26017793 \cdot 63766529 \cdot 190274191361 \cdot 1256132134125569 \cdot c_{1187}$$

and we want to find the prime factors of the remaining 1187-digit composite.

2 GNFS examples:

RSA $-$ 130 (130 digits)

$=$ 18070820886874048059516561644059055662781025167694013491
7012702145005666254024404838734112759081230337178188796
563182013214880557

$=$ 3968599945959745429016112616288837
8606757644911281006483255157243

$\times$

4553449864673597218840368689972744
08864356301263205069600999044599.

RSA $-$ 140 (140 digits)

$=$ 21290246318258757547497882016271517497806703963277721627
82333832153819499840564959113665738530219183167831073879
95317230889569230873441936471

$=$ 3398717423028438554530123627613875 8
356339864959695974234909293027714 79

$\times$

626420018740128509615165494826444 22
193020371786235090191116606539460 49.

RSA $-$ 155 (512 digits)

$=$ 10941738641570527421809707322040357612003732945449205990
91384213147634998428893478471799725789126733249762575289
9781833797076537244027146743531593354333897

$=$ 1026395928297411057720541965739916759007
16567808038066803341933521790711307779

$\times$

10660348838016845482092722036001287867920
7958575989291522270608237193062808643.

RSA − 576 (576 bits, 174 digits)

= 18819881292060796383869723946165043980716356337941738
2700763356422988859715234665485319060606505047430453173
8801130339671619969232120573403187955065699621305168
759307650257059

= 3980750864240649373971255005503864911990643
62342526708406385189575946388957261768583317

×

4727721461074353025362230719730482246329146
95302097116459852171130520711256363590397527.

RSA − 640 (193 digits, 640 bits)

= 31074182404900043721350750035888567930037346022842727
457201619488232064405180815045563468296717232867824 37
91627283803341547107310

= 16347336458092538484431338838650908598417836700 3
30923121811110852389333100104508151212118167511579

×

= 1900871281664822113126851573935413975471896789 96
85154936666385390880271038021044989571912614655 71.

RSA − 663 (200 digits, 663 bits)

= 27997833911221327870829467638722601621070446786955428
5375600099293261284001076093456710529553608560618223 5
19109513657886371059544820065767750985805576135790987
349501441788631789462951872378692218239 83

= 3532461934402770121272604978198464368671197400197 6
2502364930346877612125367942320005854795652808834 9

×

79258699544783330333470858414800596877379758573642
19960734330341455767872818152135381409304740185467.

RSA − 704 (704 bits, 212 digits)

= 74037563479561712828046796097429573142593188889231289084936232638972765034028266276891996419625117843995894330502127585370118968098286733173273108930900552505116877063299072396380786710086096962537934650563796359

= 90912135295978188784406583026004374858926083103283587204285121689604115286409333678249507883679567568061416141

× 81438592591100452657278091262844293358778990021676278832009141724293243601330041167020032408287779702524 99.

RSA − 768 (768 bits, 232 digits)

= 123018668453011775513049495838496272077285356959533479219732245215172640050726365751874520219978646938995647494277406384592519255732630345373154826850791702612214291346167042921431160222124047927473779408066535141959745985690214341 3

= 334780716989568987860441698482126908177047949837137685689124313889828837938780022876147116525317430877378144679994 89

× 367460436667995904282446337996279526322791581643430876426760322838157396665112792333734171433968102700927987363089 17.

Remark 2.3. Prior to the NFS, all modern factoring methods had an expected running time of at best

$$\mathcal{O}\left(\exp\left((c + o(1))\sqrt{\log n \log\log n}\right)\right).$$

For example, Dixon's random square method has the expected running time

$$\mathcal{O}\left(\exp\left((\sqrt{2} + o(1))\sqrt{\log n \log\log n}\right)\right),$$

whereas the MPQS takes time

$$\mathcal{O}\left(\exp\left((1+o(1))\sqrt{\log\log n/\log n}\,\right)\right).$$

Because of the Canfield–Erdős–Pomerance theorem, some people even believed that this could not be improved, except maybe for the term $(c+o(1))$, but the invention of the NFS has changed this belief.

Conjecture 2.1 (Complexity of NFS). Under some reasonable heuristic assumptions, the NFS method can factor an integer N in time

$$\mathcal{O}\left(\exp\left((c+o(1))\sqrt[3]{\log n}\sqrt[3]{(\log\log n)^2}\,\right)\right) \tag{2.11}$$

where $c = (64/9)^{1/3} \approx 1.922999427$ if GNFS is used to factor an arbitrary integer N, whereas $c = (32/9)^{1/3} \approx 1.526285657$ if SNFS is used to factor a special integer N.

ρ-Factoring Method

Although NFS is the fastest method of factoring at present, other methods are also useful, one of the particular method is the ρ-factoring method [59]; surprisingly it is the method that is applicable for all the three infeasible problems, IFP, DLP, and ECDLP discussed in this book.

ρ uses an iteration of the form

$$\left.\begin{array}{l} x_0 = \text{random}(0,\ n-1), \\[2mm] x_i \equiv f(x_{i-1}) \pmod{n}, \quad i = 1,2,3,\ldots \end{array}\right\} \tag{2.12}$$

where x_0 is a random starting value, n is the number to be factored, and $f \in \mathbb{Z}[x]$ is a polynomial with integer coefficients; usually, we just simply choose $f(x) = x^2 \pm a$ with $a \neq -2, 0$. If p is prime, then the sequence $\{x_i \bmod p\}_{i>0}$ must eventually repeat. Let $f(x) = x^2 + 1, x_0 = 0, p = 563$. Then we get the sequence $\{x_i \bmod p\}_{i>0}$ as follows (Fig. 2.3):

$$x_0 = 0,$$
$$x_1 = x_0^2 + 1 = 1,$$
$$x_2 = x_1^2 + 1 = 2,$$
$$x_3 = x_2^2 + 1 = 5,$$
$$x_4 = x_3^2 + 1 = 26,$$
$$x_5 = x_4^2 + 1 = 114,$$
$$x_6 = x_5^2 + 1 = 48,$$

$$x_7 = x_6^2 + 1 = 53,$$
$$x_8 = x_7^2 + 1 = 558,$$
$$x_9 = x_8^2 + 1 = 26.$$

That is,

$$0, 1, 2, 5, \overline{26, 114, 48, 53, 558}.$$

This sequence symbols a diagram, looks like the Greek letter ρ: As an exercise,

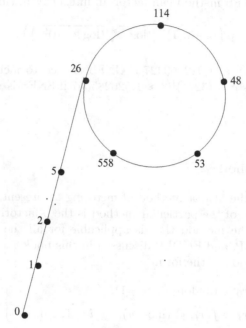

Figure 2.3. ρ cycle modulo 563 using $f(x) = x^2 + 1$ and $x_0 = 0$

readers may wish to find the ρ cycle modulo 1951 using $f(x) = x^2 + 1$ and $x_0 = 0$. Of course, to factor n, we do not know its prime factors before hand, but we can simply modulo n (justified by the Chinese Remainder Theorem). For example, to factor $n = 1098413 = 563 \cdot 1951$, we perform (all modulo 1098413):

$$x_0 = \mathbf{0},$$
$$x_1 = x_0^2 + 1 = \mathbf{1},$$

$$x_2 = x_1^2 + 1 = \mathbf{2}, \qquad y_1 = x_2 = 2 \qquad \gcd(1 - 2, n) = 1$$

$$x_3 = x_2^2 + 1 = \mathbf{5},$$

$$x_4 = x_3^2 + 1 = \mathbf{26}, \qquad y_2 = x_4 = 26 \qquad \gcd(2 - 26, n) = 1$$

$$y_i = x_{2i}$$
$$\gcd(x_i - y_i, n)$$

$$x_5 = x_4^2 + 1 = 677$$
$$\equiv 114,$$

$$x_6 = x_5^2 + 1 = 458330$$
$$\equiv 48, \qquad y_3 = x_6 = 458330 \quad \gcd(5 - 458330, n) = 1$$

$$x_7 = x_6^2 + 1 = 394716$$
$$\equiv 53,$$

$$x_8 = x_7^2 + 1 = 722324$$
$$\equiv 558, \qquad y_4 = x_8 = 722324 \quad \gcd(26 - 722324, n) = 1$$

$$x_9 = x_8^2 + 1 = 293912$$
$$\equiv 26$$

$$.x_{10} = x_9^2 + 1 = 671773$$
$$\equiv 114 \qquad y_5 = x_{10} = 671773 \quad \gcd(677 - 671773, n) = \underline{563}.$$

The following algorithm is an improved version of Brent [10] over Pollard's original ρ-method.

Algorithm 2.2 (Brent–Pollard's ρ-method). Let n be a composite integer greater than 1. This algorithm tries to find a nontrivial factor d of n, which is small compared with $\sqrt{n}$. Suppose the polynomial to use is $f(x) = x^2 + 1$.

[1] (Initialization) Choose a seed, say $x_0 = 2$, a generating function, say $f(x) = x^2 + 1 \pmod{n}$. Choose also a value for t not much bigger than $\sqrt{d}$, perhaps $t < 100\sqrt{d}$.

[2] (Iteration and Computation) Compute x_i and y_i in the following way:

$$x_1 = f(x_0),$$
$$x_2 = f(f(x_0)) = f(x_1),$$
$$x_3 = f(f(f(x_0))) = f(f(x_1)) = f(x_2),$$
$$\vdots$$
$$x_i = f(x_{i-1}).$$

$$y_1 = x_2 = f(x_1) = f(f(x_0)) = f(f(y_0)),$$
$$y_2 = x_4 = f(x_3) = f(f(x_2)) = f(f(y_1)),$$
$$y_3 = x_6 = f(x_5) = f(f(x_4)) = f(f(y_2)),$$
$$\vdots$$
$$y_i = x_{2i} = f(f(y_{i-1})).$$

and simultaneously compare x_i and y_i by computing $d = \gcd(x_i - y_i, \ n)$.

[3] (Factor Found?) If $1 < d < n$, then d is a nontrivial factor of n, print d, and go to Step [5].

[4] (Another Search?) If $x_i \equiv y_i \pmod{n}$ for some i or $i \geqslant \sqrt{t}$, then go to Step [1] to choose a new seed and a new generator and repeat.

[5] (Exit) Terminate the algorithm.

The ρ algorithm has the conjectured complexity:

Conjecture 2.2 (Complexity of the ρ-method). Let p be a prime dividing n and $p = \mathcal{O}(\sqrt{p})$, then the ρ-algorithm has expected running time

$$\mathcal{O}(\sqrt{p}) = \mathcal{O}(\sqrt{p}\,(\log n)^2) = \mathcal{O}(n^{1/4}(\log n)^2) \qquad (2.13)$$

to find the prime factor p of n.

Remark 2.4. The ρ-method is an improvement over trial division, because in trial division, $\mathcal{O}(p) = \mathcal{O}(n^{1/4})$ divisions is needed to find a small factor p of n. But of course, one disadvantage of the ρ-algorithm is that its running time is only a conjectured expected value, not a rigorous bound.

Exercises and Problems for Sect. 2.1

1. Explain why general-purpose factoring algorithms are slower than special purpose factoring algorithms, or why the special numbers are easy to factor than general numbers.

2. Show that:
   (a) Addition of two $\log n$ bit integers can be performed in $\mathcal{O}(\log n)$ bit operations.

   (b) Multiplication of two $\log n$ bit integers can be performed in $\mathcal{O}((\log n)^{1+\epsilon})$ bit operations.

3. Show that:
   (a) Assume the ERH, there is deterministic algorithm that factors n in $\mathcal{O}(n^{1/5+\epsilon})$ steps.

   (b) FFT (fast Fourier transform) can be utilized to factor an integer n in $\mathcal{O}(n^{1/4+\epsilon})$ steps.

   (c) Give two deterministic algorithms that factor integer n in $\mathcal{O}(n^{1/3+\epsilon})$ steps.

4. Show that if $\mathcal{P} = \mathcal{NP}$, then IFP $\in \mathcal{P}$.

5. Prove or disprove that IFP $\in \mathcal{NP}$-complete.

6. Extend the NFS to FFS (function field sieve). Give a complete description of the FFS for factoring large integers.

7. Let $x_i = f(x_{i-1})$, $i = 1, 2, 3, \ldots$. Let also $t, u > 0$ be the smallest numbers in the sequence $x_{t+i} = x_{t+u+i}$, $i = 0, 1, 2, \ldots$, where t and u are called the lengths of the ρ tail and cycle, respectively. Give an efficient algorithm to determine t and u exactly, and analyze the running time of your algorithm.

8. Find the prime factorization of the following RSA numbers, each of these numbers has two prime factors:

(a) RSA-896 (270 digits, 896 bits)
412023436986659543855531365332575948179811699844327982845455626433876445565248426198098870423161841879261420247188869492560931776375033342113098239748515094490910691026986103186270411488086697056490290365365886743373172081310410519086425479328260139125762403394637326939 1

(b) RSA-1024 (309 digits, 1024 bits)
135066410865995223349603216278805969938881475605667027524485143851526510604859533833940287150571909441798207282164471551373680419703964191743046496589274256239341020864383202110372958725762358509643110564073501508187510676594629205563685529475213500852879416377328533906109750544334999811150056977236890927563

(c) RSA-1536 (463 digits, 1536 bits)
184769970321174147430683356202001644030185493386634101714717857749106516967111612498593376843054357445856160615445717940522297177325246609606469460712496237204420222697567566873784275623895087646784409332851574965788434150884755282981867264513398633649319080846719904318743812833635027954702826532978029349161558118810498449083195450009848393775227257052578591944993870073695755688436933812779613089230392569695253261620823676490316036553137144791393234716956698806 9

(d) RSA-2048 (617 digits, 2048 bits)
251959084756578934940271832400483985714292821262040320277771378360436620207075555626401852588078440691829064124951508218929855914917618450280848912007284499268739280728777673597141834727026189637501497182469116507761337985909570009733045974880842840179742910064245869181719511874612151517265463228221686998754918242243363725908514186546204357679842338718477444792073993423658482382428119816381501067481045166037730605620161967625613384414360383390441495263443219011465754445417842402092461651572335077870774981712577246796292638635637328991215483 14

38167899885040445364023527381951378636564391212010397122822120720357

9. Try to complete the following prime factorization of the smallest unfactored (not completely factored) Fermat numbers:

$$F_{12} = 114689 \cdot 26017793 \cdot 63766529 \cdot 190274191361 \cdot \\ 1256132134125569 \cdot c_{1187}$$

$$F_{13} = 2710954639361 \cdot 2663848877152141313 \cdot 36031098445229199 \cdot \\ 319546020820551643220672513 \cdot c_{2391}$$

$$F_{14} = c_{4933}$$

$$F_{15} = 1214251009 \cdot 2327042503868417 \cdot \\ 168768817029516972383024127016961 \cdot c_{9808}$$

$$F_{16} = 825753601 \cdot 188981757975021318420037633 \cdot c_{19694}$$

$$F_{17} = 31065037602817 \cdot c_{39444}$$

$$F_{18} = 13631489 \cdot 81274690703860512587777 \cdot c_{78884}$$

$$F_{19} = 70525124609 \cdot 646730219521 \cdot c_{157804}$$

$$F_{20} = c_{315653}$$

$$F_{21} = 4485296422913 \cdot c_{631294}$$

$$F_{22} = c_{1262612}$$

$$F_{23} = 167772161 \cdot c_{2525215}$$

$$F_{24} = c_{5050446}$$

Basically, you are asked to factor the unfactored composite numbers, denoted by c_x, of the Fermat numbers. For example, in F_{12}, c_{1187} is the unfactored 1187 digit composite.

2.2 IFP-Based Cryptography

Basic Idea of IFP-Based Cryptography

IFP-based cryptography is a class of cryptographic systems whose security relies on the intractability of the IFP problem:

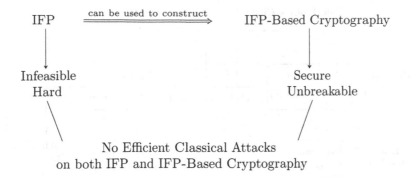

No Efficient Classical Attacks
on both IFP and IFP-Based Cryptography

Typical cryptographic systems in this class include RSA [64], Rabin [62], and Goldwasser–Micali probabilistic encryption [32] and Goldwasser–Micali–Rackoff zero-knowledge interactive proof [33]. We shall first give an account of the RSA cryptographic system. In a general cryptographic setting, we assume Alice wishes to send a ciphertext C of the plaintext M to Bob (or vice versa), Eve, the eavesdropper, wishes to understand the communication between Alice and Bob:

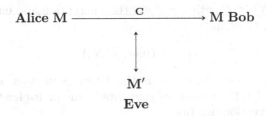

RSA Cryptography

RSA is the most famous, first practical, widely used, and still unbreakable public-key cryptography, for which its three inventors, Rivest, Shamir, and Adleman, received the 2002 Turing Award. The security of RSA relies completely on the infeasibility of the IFP problem.

Definition 2.3. The *RSA public-key cryptosystem* may be formally defined as follows (Depicted in Fig. 2.4):

$$\text{RSA} = (\mathcal{M}, \mathcal{C}, \mathcal{K}, M, C, e, d, N, E, D) \tag{2.14}$$

where:

1. $\mathcal{M}$ is the set of plaintexts, called the plaintext space.
2. $\mathcal{C}$ is the set of cipherexts, called the ciphertext space.
3. $\mathcal{K}$ is the set of keys, called the key space.
4. $M \in \mathcal{M}$ is a piece of particular plaintext.

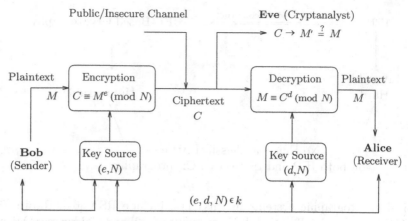

Figure 2.4. RSA public-key cryptography

5. $C \in \mathcal{C}$ is a piece of particular ciphertext.

6. $N = pq$ is the modulus with p, q prime numbers, usually each with at least 100 digits.

7. $\{(e, N), (d, N)\} \in \mathcal{K}$ with $e \neq d$ are the encryption and encryption keys, respectively, satisfying

$$ed \equiv 1 \ (\text{mod} \ \phi(N)) \tag{2.15}$$

where $\phi(N) = (p-1)(q-1)$ is the Euler ϕ-function and defined by $\phi(N) = \#(\mathbb{Z}_N^*)$, the number of elements in the multiplicative group $\mathbb{Z}_N^*$.

8. E is the encryption function

$$E_{e,N} : \ M \mapsto C$$

That is, $M \in \mathcal{M}$ maps to $C \in \mathcal{C}$, using the public-key (e, N), such that

$$C \equiv M^e \ (\text{mod} \ N). \tag{2.16}$$

9. D is the decryption function

$$D_{d,N} : \ C \mapsto M$$

That is, $C \in \mathcal{C}$ maps to $M \in \mathcal{M}$, using the private-key (d, N), such that

$$M \equiv C^d \equiv (M^e)^d \ (\text{mod} \ N). \tag{2.17}$$

The idea of RSA can be best depicted in Fig. 2.5.

Theorem 2.3 (The Correctness of RSA). Let M, C, N, e, d be plaintext, ciphertext, encryption exponent, decryption exponent, and modulus, respectively. Then

$$(M^e)^d \equiv M \ (\text{mod} \ N).$$

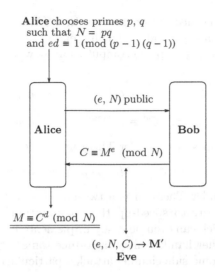

Alice chooses primes p, q
such that $N = pq$
and $ed \equiv 1 \,(\mathrm{mod}\, (p-1)(q-1))$

(e, N) public

Alice

Bob

$C \equiv M^e \;(\mathrm{mod}\; N)$

$M \equiv C^d \;(\mathrm{mod}\; N)$

$(e, N, C) \to \mathbf{M'}$
Eve

Figure 2.5. RSA encryption and decryption

Proof. Notice first that

$$
\begin{aligned}
C^d &\equiv (M^e)^d \;(\mathrm{mod}\; N) && (\text{since } C \equiv M^e \;(\mathrm{mod}\; N))\\
&\equiv M^{1+k\phi(N)} \;(\mathrm{mod}\; N) && (\text{since } ed \equiv 1 \;(\mathrm{mod}\; \phi(N)))\\
&\equiv M \cdot M^{k\phi(N)} \;(\mathrm{mod}\; N)\\
&\equiv M \cdot (M^{\phi(N)})^k \;(\mathrm{mod}\; N)\\
&\equiv M \cdot (1)^k \;(\mathrm{mod}\; N) && (\text{by Euler's Theorem } a^{\phi(n)} \equiv 1 \;(\mathrm{mod}\; N))\\
&\equiv M
\end{aligned}
$$

The result thus follows. □

Both encryption $C \equiv M^e \;(\mathrm{mod}\; N)$ and decryption $M \equiv C^d \;(\mathrm{mod}\; N)$ of RSA can be implemented in polynomial time by the fast exponentiation method. For example, the RSA encryption can be implemented as follows:

Algorithm 2.3. Given (e, M, N), this algorithm finds $C \equiv M^e \;(\mathrm{mod}\; N)$, or given (d, C, N), finds $M \equiv C^d \;(\mathrm{mod}\; N)$ in time polynomial in $\log e$ or $\log d$, respectively.

Encryption:
Given (e, M, N) to find C
Set $C \leftarrow 1$
While $e \geqslant 1$ do
    if $e \bmod 2 = 1$
        then $C \leftarrow C \cdot M \bmod N$
    $M \leftarrow M^2 \bmod N$
    $e \leftarrow \lfloor e/2 \rfloor$
Print C

Decryption:
Given (d, C, N) to find M
Set $M \leftarrow 1$
    While $d \geqslant 1$ do
        if $d \bmod 2 = 1$
           then $M \leftarrow M \cdot C \bmod N$
        $C \leftarrow C^2 \bmod N$
        $d \leftarrow \lfloor d/2 \rfloor$
    Print M

Remark 2.5. For the decryption process in RSA, as the authorized user knows d and hence knows p and q, thus instead of directly working on $M \equiv C^d \pmod{N}$, he can speed-up the computation by working on the following two congruences:

$$M_p \equiv C^d \equiv C^{d \bmod p-1} \pmod{p}$$

$$M_q \equiv C^d \equiv C^{d \bmod q-1} \pmod{q}$$

and then use the Chinese Remainder Theorem to get

$$M \equiv M_p \cdot q \cdot q^{-1} \bmod p + M_q \cdot p \cdot p^{-1} \bmod q \pmod{N}. \qquad (2.18)$$

The Chinese Remainder Theorem is a two-edged sword. On the one hand, it provides a good way to speed-up the computation/performance of the RSA decryption, which can even be easily implemented by a low-cost crypto-chip [34]. On the other hand, it may introduce some serious security problems vulnerable to some side-channel attacks, particularly the random fault attacks;

Example 2.6. Let the letter-digit encoding be as follows:

$$\text{space} = 00, A = 01, B = 02, \cdots, Z = 26.$$

(We will use this digital representation of letters throughout the book.) Let also

$$e = 9007,$$

$$M = 20080500130107090300231518041900011805001917210501 1309\_$$
$$1908001519190906180 10705,$$

$$N = 11438162575788886766923577997614661201021829672124 2362\_$$
$$56256184293570693524573389783059712356395870505898 9075\_$$
$$147599290026879543541.$$

Then the encryption can be done by using Algorithm 2.3:

$$C \equiv M^e$$

$$\equiv 96869613754622061477140922254355882905759991124574 3198\_$$
$$74695120930816298225145708356931476622883989628013 3919\_$$
$$9055182994515781 5154 \pmod{N}.$$

For the decryption, since the two prime factors p and q of N are known to the authorized person who does the decryption:

$$p = 34905295108476509491478496199038981334177646384933 878\_$$
$$43990820577$$

$$q = 32769132993266709549961988190834461413177642967992 942\_$$
$$539798288533$$

then

$$
\begin{aligned}
d &\equiv 1/e \\
&\equiv 106698614368578024442868771328920154780709906633937862\_ \\
&\equiv 80122622449663106312591177447087334016859746230655396\_8 \\
&\equiv 544513277109053606095 \ (\mathrm{mod} \ (p-1)(q-1)).
\end{aligned}
$$

Thus, the original plaintext M can be recovered either directly by using Algorithm 2.3 or indirectly by a combined use of Algorithm 2.3 and the Chinese Remainder Theorem (2.18):

$$
\begin{aligned}
M &\equiv C^d \\
&= 200805001301070903002315180419000118050019172105011309\_ \\
&\quad 190800151919090618010705 \ (\mathrm{mod} \ N)
\end{aligned}
$$

which is "THE MAGIC WORDS ARE SQUEAMISH OSSIFRAGE."

Remark 2.6. Prior to RSA, Pohlig and Hellman in 1978 [57] proposed a secret-key cryptography based on arithmetic modulo p, rather than $N = pq$. The Pohlig–Hellman system works as follows: Let M and C be the plain and cipher texts, respectively. Choose a prime p, usually with more than 200 digits, and a secret encryption key e such that $e \in \mathbb{Z}^+$ and $e \leqslant p-2$. Compute $d \equiv 1/e \ (\mathrm{mod} \ (p-1))$. (e, p) and of course d must be kept as a secret.

[1] **Encryption:**

$$
C \equiv M^e \ (\mathrm{mod} \ p). \tag{2.19}
$$

This process is easy for the authorized user:

$$
\{M, e, p\} \xrightarrow[\text{easy}]{\text{find}} \{C \equiv M^e \ (\mathrm{mod} \ p)\}. \tag{2.20}
$$

[2] **Decryption:**

$$
M \equiv C^d \ (\mathrm{mod} \ p). \tag{2.21}
$$

For the authorized user who knows (e, p), this process is easy, since d can be easily computed from e.

[3] **Cryptanalysis:** The security of this system is based on the infeasibility of the discrete logarithm problem. For example, for a cryptanalyst who does not know e or d would have to compute:

$$
e \equiv \log_M C \ (\mathrm{mod} \ p).
$$

Remark 2.7. One of the most important features of RSA encryption is that it can also be used for digital signatures. Let M be a document to be signed, and $N = pq$ with p, q primes, (e, d) the public and private exponents as in RSA encryption scheme. Then the processes of RSA signature signing and signature verification are just the same as that of the decryption and encryption; that is, use d for signature signing and e signature verification as follows (see also Fig. 2.6):

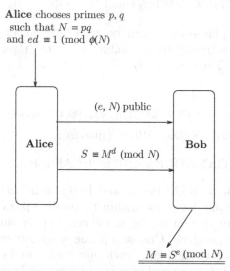

Figure 2.6. RSA digital signature

[1] **Signature signing**:

$$S \equiv M^d \pmod{N} \tag{2.22}$$

The signing process can only be done by the authorized person who has the private exponent d.

[2] **Signature verification**:

$$M \equiv S^e \pmod{N} \tag{2.23}$$

This verification process can be done by anyone since (e, N) is public.

Of course, RSA encryption and RSA signature can be used together to obtain a signed encrypted document to be sent over an insecure network.

RSA Problem and RSA Assumption

As can be seen from the previous section, the whole idea of the RSA encryption and decryption is as follows:

$$\left. \begin{array}{rcl} C & \equiv & M^e \pmod{N}, \\ M & \equiv & C^d \pmod{N} \end{array} \right\} \tag{2.24}$$

where

$$\left. \begin{array}{rcl} ed & \equiv & 1 \ (\mathrm{mod} \ \phi(N)) \\ N & = & pq \ \text{with} \ p, q \in \text{Primes.} \end{array} \right\} \qquad (2.25)$$

Thus, the *RSA function* can be defined by

$$f_{\mathrm{RSA}} : M \mapsto M^e \ \mathrm{mod} \ N. \qquad (2.26)$$

The *inverse of the RSA function* is then defined by

$$f_{\mathrm{RSA}}^{-1} : M^e \mapsto M \ \mathrm{mod} \ N. \qquad (2.27)$$

Clearly, the RSA function is a *one-way trap-door function*, with

$$\{d, p, q, \phi(N)\} \qquad (2.28)$$

the RSA *trap-door information*mitrap-door information. For security purposes, this set of information must be kept as a secret and should never be disclosed in anyway even in part. Now, suppose that Bob sends C to Alice, but Eve intercepts it and wants to understand it. Since Eve only has (e, N, C) and does not have any piece of the trap-door information in (2.28), then it should be infeasible/intractable for her to recover M from C:

$$\{e, N, C = M^e \ (\mathrm{mod} \ N)\} \xrightarrow{\text{hard}} \{M \equiv C^d \ (\mathrm{mod} \ N)\}. \qquad (2.29)$$

On the other hand, for Alice, since she knows d, which implies that she knows all the pieces of trap-door information in (2.28), since

$$\{d\} \xleftrightarrow{\mathcal{P}} \{p\} \xleftrightarrow{\mathcal{P}} \{q\} \xleftrightarrow{\mathcal{P}} \{\phi(N)\} \qquad (2.30)$$

Thus, it is easy for Alice to recover M from C:

$$\{N, C \equiv M^e \ (\mathrm{mod} \ N)\} \xrightarrow[\text{easy}]{\{d,p,q,\phi(N)\}} \{M \equiv C^d \ (\mathrm{mod} \ N)\}. \qquad (2.31)$$

Why is it hard for Eve to recover M from C? This is because Eve is facing a hard computational problem, namely, the *RSA problem* [65]:

The RSA problem: Given the RSA public-key (e, N) and the RSA ciphertext C, find the corresponding RSA plaintext M. That is,

$$\{e, N, C\} \longrightarrow \{M\}.$$

It is conjectured although it has never been proved or disproved that:

The RSA conjecture: Given the RSA public-key (e, N) and the RSA ciphertext C, it is hard to find the corresponding RSA plaintext M. That is,

$$\{e, N, C\} \xrightarrow{\text{hard}} \{M\}.$$

But how hard is it for Alice to recover M from C? This is another version of the RSA conjecture, often called the *RSA assumption*, which again has never been proved or disproved:

> **The RSA assumption:** Given the RSA public-key (e, N) and the RSA ciphertext C, then finding M is as hard as factoring the RSA modulus N. That is,
>
> $$\text{IFP}(N) \iff \text{RSA}(M)$$
>
> provided that N is sufficiently large and randomly generated, and M and C are random integers between 0 and $N - 1$. More precisely, it is conjectured (or assumed) that
>
> $$\text{IFP}(N) \overset{\mathcal{P}}{\iff} \text{RSA}(M).$$

That is, if N can be factorized in polynomial time, then M can be recovered from C in polynomial time. In other words, cryptoanalyzing RSA must be as difficult as solving the IFP problem. But the problem is, as we discussed previously, that no one knows whether or not IFP can be solved in polynomial time, so RSA is only assumed to be secure, not proved to be secure:

$$\text{IFP}(N) \text{ is hard} \longrightarrow \text{RSA}(M) \text{ is secure.}$$

The real situation is that

$$\text{IFP}(N) \overset{\checkmark}{\Longrightarrow} \text{RSA}(M),$$

$$\text{IFP}(N) \overset{?}{\Longleftarrow} \text{RSA}(M).$$

Now, we can return to answer the question of how hard is it for Alice to recover M from C? By the RSA assumption, cryptanalyzing C is as hard as factoring N. The fastest known integer factorization algorithm, the NFS, runs in time

$$\mathcal{O}(\exp(c(\log N)^{1/3}(\log \log N)^{2/3}))$$

where $c = (64/9)^{1/3}$ if a general version of NFS, GNFS, is used for factoring an arbitrary integer N whereas $c = (32/9)^{1/3}$ if a special version of NFS, SNFS, is used for factoring a special form of integer N. As in RSA, the modulus $N = pq$ is often chosen to be a large general composite integer $N = pq$ with p and q of the same bit size, which makes SNFS not useful. This means that RSA cannot be broken in polynomial time but in subexponential-time, which makes RSA secure, again, by assumption. Thus, readers should note that the RSA problem is *assumed* to be *hard*, and the RSA cryptosystem is *conjectured* to be *secure* .

In the RSA cryptosystem, it is assumed that the cryptanalyst, Eve:

1. Knows the public-key $\{e, N\}$ with $N = pq$ and also the ciphertext C
2. Does not know any one piece of the trap-door information $\{p, q, \phi(N), d\}$
3. Wants to know $\{M\}$

That is,

$$\{e, N, C \equiv M^e \pmod{N}\} \xrightarrow{\text{Eve wants to find}} \{M\}.$$

Obviously, there are several ways to recover M from C (i.e., to break the RSA system):

1. Factor N to get $\{p, q\}$ so as to compute

$$M \equiv C^{1/e \ (\bmod \ (p-1)(q-1))} \pmod{N}.$$

2. Find $\phi(N)$ so as to compute

$$M \equiv C^{1/e \ (\bmod \ \phi(N))} \pmod{N}.$$

3. Find order(a, N), the order of a random integer $a \in [2, N-2]$ modulo N, then try to find

$$\{p, q\} = \gcd(a^{r/2} \pm 1, N) \text{ and } M \equiv C^{1/e \ (\bmod \ (p-1)(q-1))} \pmod{N}.$$

4. Find order(C, N), the order of C modulo N, so as to compute

$$M \equiv C^{1/e \ (\bmod \ \text{order}(C,N))} \pmod{N}.$$

5. Compute $\log_C M \pmod{N}$, the discrete logarithm M to the base C modulo N in order to find

$$M \equiv C^{\log_C M \ (\bmod \ N)} \pmod{N}$$

Rabin Cryptography

As can be seen from the previous sections, RSA uses M^e for encryption, with $e \geqslant 3$ (3 is the smallest possible public exponent in RSA); in this way, we might call RSA encryption M^e encryption. In 1979, Michael Rabin [62] proposed a scheme based on M^2 encryption, rather than the M^e for $e \geqslant 3$ encryption used in RSA. A brief description of the Rabin cryptosystem is as follows (see also Fig. 2.7).

1. **Key generation:** Let $n = pq$ with p, q odd primes satisfying

$$p \equiv q \equiv 3 \pmod{4}. \tag{2.32}$$

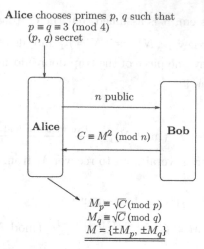

Figure 2.7. Rabin cryptosystem

2. **Encryption:**

$$C \equiv M^2 \ (\text{mod } n).\tag{2.33}$$

3. **Decryption:** Use the Chinese Remainder Theorem to solve the system of congruences:

$$\begin{cases} M_p \equiv \sqrt{C} \ (\text{mod } p) \\ M_q \equiv \sqrt{C} \ (\text{mod } q) \end{cases}\tag{2.34}$$

to get the four solutions: $\{\pm M_p, \pm M_q\}$. The true plaintext M will be one of these four values.

4. **Cryptanalysis:** A cryptanalyst who can factor n can compute the four square roots of C modulo n and hence can recover M from C. Thus, breaking the Rabin system is equivalent to factoring n.

Example 2.7. Let $M = 31$.

[1] **Key generation:** Let $n = 11 \cdot 19$ be the public-key, but keep the prime factors $p = 11$ and $q = 19$ of n as a secret.

[2] **Encryption:**

$$C \equiv 31^2 \equiv 125 \ (\text{mod } 209).$$

[3] **Decryption:** Compute

$$\begin{cases} M_p \equiv \sqrt{125} \equiv \pm 2 \ (\text{mod } p) \\ M_q \equiv \sqrt{125} \equiv \pm 7 \ (\text{mod } q). \end{cases}$$

Now, use the Chinese Remainder Theorem to solve

$$\begin{cases} M \equiv 2 \ (\mathrm{mod}\ 11) \\ M \equiv 7 \ (\mathrm{mod}\ 19) \end{cases} \Longrightarrow M = 178$$

$$\begin{cases} M \equiv -2 \ (\mathrm{mod}\ 11) \\ M \equiv 7 \ (\mathrm{mod}\ 19) \end{cases} \Longrightarrow M = 64$$

$$\begin{cases} M \equiv -2 \ (\mathrm{mod}\ 11) \\ M \equiv 7 \ (\mathrm{mod}\ 19) \end{cases} \Longrightarrow M = 145$$

$$\begin{cases} M \equiv -2 \ (\mathrm{mod}\ 11) \\ M \equiv -7 \ (\mathrm{mod}\ 19) \end{cases} \Longrightarrow M = 31$$

The true plaintext M will be one of the above four values, and in fact, $M = 31$ is the true value.

Unlike the RSA cryptosystem whose security was only conjectured to be equivalent to the intractability of IFP, the security of Rabin system and its variant such as Rabin–Williams system is proved to be equivalent to the intractability of IFP. First, notice that there is a fast algorithm to compute the square roots modulo N if $n = pq$ is known. Consider the following quadratic congruence

$$x^2 \equiv y \ (\mathrm{mod}\ p) \tag{2.35}$$

there are essentially three cases for the prime p:

(1) $p \equiv 3 \ (\mathrm{mod}\ 4)$.

(2) $p \equiv 5 \ (\mathrm{mod}\ 8)$.

(3) $p \equiv 1 \ (\mathrm{mod}\ 8)$.

All three cases may be solved by the following process:

$$\begin{cases} \text{if } p \equiv 3 \ (\mathrm{mod}\ 4), \quad x \equiv \pm y^{\frac{p+1}{4}} \ (\mathrm{mod}\ p), \\[2mm] \text{if } p \equiv 5 \ (\mathrm{mod}\ 8), \quad \begin{cases} \text{if } y^{\frac{p+1}{4}} = 1, \quad x \equiv \pm y^{\frac{p+3}{8}} \ (\mathrm{mod}\ p) \\[2mm] \text{if } y^{\frac{p+1}{4}} \neq 1, \quad x \equiv \pm 2y(4y)^{\frac{p-5}{8}} \ (\mathrm{mod}\ p). \end{cases} \end{cases} \tag{2.36}$$

Algorithm 2.4 (Computing square roots modulo pq). Let $n = pq$ with p and q odd prime and $y \in \mathrm{QR}_n$. This algorithm will find all the four solutions in x to congruence $x^2 \equiv y \ (\mathrm{mod}\ pq)$ in time $\mathcal{O}((\log p)^4)$.

[1] Use (2.36) to find a solution r to $x^2 \equiv y \ (\mathrm{mod}\ p)$.

[2] Use (2.36) to find a solution s to $x^2 \equiv y \pmod{q}$.

[3] Use the Extended Euclid's algorithm to find integers c and d such that $cp + dq = 1$.

[4] Compute $x \equiv \pm(rdq \pm scp) \pmod{pq}$.

On the other hand, if there exists an algorithm to find the four solutions in x to $x^2 \equiv y \pmod{n}$, then there exists an algorithm to find the prime factorization of n. The following is the algorithm.

Algorithm 2.5 (Factoring via square roots). This algorithm seeks to find a factor of n by using an existing square root finding algorithm (namely, Algorithm 2.4).

[1] Choose at random an integer x such that $\gcd(x, n) = 1$, and compute $x^2 \equiv a \pmod{n}$.

[2] Use Algorithm 2.4 to find four solutions in x to $x^2 \equiv a \pmod{n}$.

[3] Choose one of the four solutions, say y such that $y \not\equiv \pm x \pmod{n}$, then compute $\gcd(x \pm y, n)$.

[4] If $\gcd(x \pm y, n)$ reveals p or q, then go to Step [5], or otherwise, go to Step [1].

[5] Exit.

Theorem 2.4. Let $N = pq$ with p, q odd prime. If there exists a polynomial-time algorithm A to factor $n = pq$, then there exists an algorithm B to find a solution to $x^2 \equiv y \pmod{n}$, for any $y \in \mathrm{QR}_N$.

Proof. If there exists an algorithm A to factor $n = pq$, then there exists an algorithm (in fact, Algorithm 2.4), which determines $x = \pm(rdq \pm scp) \pmod{pq}$, as defined in Algorithm 2.4, for $x^2 \equiv y \pmod{n}$. Clearly, Algorithm 2.4 runs in polynomial time.   □

Theorem 2.5. Let $n = pq$ with p, q odd prime. If there exists a polynomial-time algorithm A to find a solution to $x^2 \equiv a \pmod{n}$, for any $a \in \mathrm{QR}_n$, then there exists a probabilistic polynomial-time algorithm B to find a factor of n.

Proof. First, note that for n composite, x and y integer, if $x^2 \equiv y^2 \pmod{n}$ but $x \not\equiv \pm y \pmod{n}$, then $\gcd(x + y, n)$ are proper factors of n. If there exists an algorithm A to find a solution to $x^2 \equiv a \pmod{n}$ for any $a \in \mathrm{QR}_n$, then there exists an algorithm (in fact, Algorithm 2.5), which uses algorithm A to find four solutions in x to $x^2 \equiv a \pmod{n}$ for a random x with $\gcd(x, n) = 1$. Select one of the solutions, say, $y \not\equiv \pm x \pmod{n}$, then by computing $\gcd(x \pm y, n)$, the probability of finding a factor of N will be $\geqslant 1/2$. If Algorithm 2.5 runs for k times and each time randomly chooses a different x, then the probability of not factoring n is $\leqslant 1/2^k$.   □

So, finally, we have

Theorem 2.6. Factoring integers, computing the modular square roots, and breaking the Rabin cryptosystem are computationally equivalent. That is,

$$\mathrm{IFP}(n) \overset{\mathcal{P}}{\Longleftrightarrow} \mathrm{Rabin}(M). \tag{2.37}$$

Residuosity-Based Cryptography

Recall that an integer a is a quadratic residue modulo n, denoted by $a \in Q_n$, if $\gcd(a, n) = 1$ and there exists a solution x to the congruence $x^2 \equiv a \pmod{n}$, otherwise a is a quadratic non-residue modulo n, denoted by $a \in \overline{Q}_n$. The quadratic residuosity problem (QRP) may be stated as:

Given positive integers a and n, decide whether or not $a \in Q_n$.

It is believed that solving QRP is equivalent to computing the prime factorization of n, so it is computationally infeasible. If n is prime then

$$a \in Q_n \quad \Longleftrightarrow \quad \left(\frac{a}{n}\right) = 1, \tag{2.38}$$

and if n is composite, then

$$a \in Q_n \quad \Longrightarrow \quad \left(\frac{a}{n}\right) = 1, \tag{2.39}$$

but

$$a \in Q_n \quad \overset{\times}{\Longleftarrow} \quad \left(\frac{a}{n}\right) = 1, \tag{2.40}$$

however,

$$a \in \overline{Q}_n \quad \Longleftarrow \quad \left(\frac{a}{n}\right) = -1. \tag{2.41}$$

Let $J_n = \{a \in (\mathbb{Z}/n\mathbb{Z})^* : \left(\frac{a}{n}\right) = 1\}$, then $\tilde{Q}_n = J_n - Q_n$. Thus, $\tilde{Q}_n$ is the set of all pseudosquares modulo n; it contains those elements of J_n that do not belong to Q_n. Readers may wish to compare this result to Fermat's little theorem, namely (assuming $\gcd(a, n) = 1$),

$$n \text{ is prime} \quad \Longrightarrow \quad a^{n-1} \equiv 1 \pmod{n}, \tag{2.42}$$

but

$$n \text{ is prime} \quad \overset{\times}{\Longleftarrow} \quad a^{n-1} \equiv 1 \pmod{n}, \tag{2.43}$$

however,

$$n \text{ is composite} \quad \Longleftarrow \quad a^{n-1} \not\equiv 1 \pmod{n}. \tag{2.44}$$

The QRP can then be further restricted to:

Given a composite n and an integer $a \in J_n$, decide whether or not $a \in Q_n$.

For example, when $n = 21$, we have $J_{21} = \{1, 4, 5, 16, 17, 20\}$ and $Q_{21} = \{1, 4, 16\}$, thus $\tilde{Q}_{21} = \{5, 17, 20\}$. So, the QRP problem for $n = 21$ is actually to distinguish squares $\{1, 4, 16\}$ from pseudosquares $\{5, 17, 20\}$. The only method we know for distinguishing squares from pseudosquares is to factor n; since integer factorization is computationally infeasible, the QRP problem is computationally infeasible. In what follows, we shall present a cryptosystem whose security is based on the infeasibility of the QRP; it was first proposed by Goldwasser and Micali in 1984 [32] in 1984, under the term *probabilistic encryption*.

Algorithm 2.6 (Quadratic residuosity-based cryptography). This algorithm uses the randomized method to encrypt messages and is based on the QRP. The algorithm divides into three parts: key generation, message encryption, and decryption.

[1] Key generation: Both Alice and Bob should do the following to generate their public and secret keys:

    [a] Select two large distinct primes p and q, each with roughly the same size, say, each with β bits.

    [b] Compute $n = pq$.

    Select a $y \in \mathbb{Z}/n\mathbb{Z}$, such that $y \in \overline{Q}_n$ and $\left(\dfrac{y}{n}\right) = 1$. ($y$ is thus a pseudosquare modulo n).

    [c] Make (n, y) public, but keep (p, q) secret.

[2] Encryption: To send a message to Alice, Bob should do the following:
    [a] Obtain Alice's public-key (n, y).

    [c] Represent the message m as a binary string $m = m_1 m_2 \cdots m_k$ of length k.

    [d] For i from 1 to k do
        [d-1] Choose at random an $x \in (\mathbb{Z}/n\mathbb{Z})^*$ and call it x_i.

        [d-2] Compute c_i:

$$c_i = \begin{cases} x_i^2 \bmod n, & \text{if } m_i = 0, \quad \text{(r.s.)} \\ y x_i^2 \bmod n, & \text{if } m_i = 1, \quad \text{(r.p.s.)}, \end{cases} \tag{2.45}$$

        where r.s. and r.p.s. represent random square and random pseudosquare, respectively.

    Send the k-tuple $c = (c_1, c_2, \ldots, c_k)$ to Alice. (Note first that each c_i is an integer with $1 \leqslant c_i < n$. Note also that since n is a 2β-bit integer, it is clear that the cipher-text c is a much longer string than the original plain-text m.)

[3] Decryption: To decrypt Bob's message, Alice should do the following:

[a] For i from 1 to k do

[a-1] Evaluate the Legendre symbol:

$$e'_i = \left(\frac{c_i}{p} \right). \qquad (2.46)$$

[a-2] Compute m_i:

$$m_i = \begin{cases} 0, & \text{if } e'_i = 1 \\ 1, & \text{if otherwise.} \end{cases} \qquad (2.47)$$

That is, $m_i = 0$ if $c_i \in Q_n$, otherwise, $m_i = 1$.

Finally, get the decrypted message $m = m_1 m_2 \cdots m_k$.

Remark 2.8. The above encryption scheme has the following interesting features:

1) The encryption is random in the sense that the same bit is transformed into different strings depending on the choice of the random number x. For this reason, it is called *probabilistic* (or *randomized*) encryption.

2) Each bit is encrypted as an integer modulo n and hence is transformed into a 2β-bit string.

3) It is semantically secure against any threat from a polynomially bounded attacker, provided that the QRP is hard.

Example 2.8. In what follows we shall give an example of how Bob can send the message "HELP ME" to Alice using the above cryptographic method. We use the binary equivalents of letters as defined in Table 2.1. Now, both Alice

Table 2.1. The binary equivalents of letters

Letter	Binary code	Letter	Binary code	Letter	Binary code
A	00000	B	00001	C	00010
D	00011	E	00100	F	00101
G	00110	H	00111	I	01000
J	01001	K	01010	L	01011
J	01001	K	01010	L	01011
M	01100	N	01101	O	01110
P	01111	Q	10000	R	10001
S	10010	T	10011	U	10100
V	10101	W	10110	X	10111
Y	11000	Z	11001	⊔	11010

and Bob proceed as follows:

[1] Key generation:
- Alice chooses $(n, y) = (21, 17)$ as a public-key, where $n = 21 = 3 \cdot 7$ is a composite and $y = 17 \in \tilde{Q}_{21}$ (since $17 \in J_{21}$ but $17 \notin Q_{21}$), so that Bob can use the public-key to encrypt his message and send it to Alice.
- Alice keeps the prime factorization $(3, 7)$ of 21 as a secret; since $(3, 7)$ will be used as a private decryption key. (Of course, here we just show an example; in practice, the prime factors p and q should be at last 100 digits.)

[2] Encryption:
- Bob converts his plaintext HELP ME to the binary stream $M = m_1 m_2 \cdots m_{35}$:

$$00111\ 00100\ 01011\ 01111\ 11010\ 01100\ 00100.$$

(To save space, we only consider how to encrypt and decrypt $m_2 = 0$ and $m_3 = 1$; readers are suggested to encrypt and decrypt the whole binary stream.)

- Bob randomly chooses integers $x_i \in (\mathbb{Z}/21\mathbb{Z})^*$. Suppose he chooses $x_2 = 10$ and $x_3 = 19$ which are elements of $(\mathbb{Z}/21\mathbb{Z})^*$.

- Bob computes the encrypted message $C = c_1 c_2 \cdots c_k$ from the plaintext $M = m_1 m_2 \cdots m_k$ using (2.45). To get, for example, c_2 and c_3, Bob performs:

$$c_2 = x_2^2 \bmod 21 = 10^2 \bmod 21 = 16, \qquad \text{since } m_2 = 0,$$
$$c_3 = y \cdot x_3^2 \bmod 21 = 17 \cdot 19^2 \bmod 21 = 5, \quad \text{since } m_3 = 1.$$

(Note that each c_i is an integer reduced to 21, i.e., m_i is a bit, but its corresponding c_i is not a bit but an integer, which is a string of bits, determined by Table 2.1.)

- Bob then sends c_2 and c_3 along with all other c_i's to Alice.

[3] Decryption: To decrypt Bob's message, Alice evaluates the Legendre symbols $\left(\dfrac{c_i}{p}\right)$ and $\left(\dfrac{c_i}{q}\right)$. Since Alice knows the prime factorization (p, q) of n, it should be easy for her to evaluate these Legendre symbols. For example, for c_2 and c_3, Alice first evaluates the Legendre symbols $\left(\dfrac{c_i}{p}\right)$:

$$e_2' = \left(\frac{c_2}{p}\right) = \left(\frac{16}{3}\right) = \left(\frac{4^2}{3}\right) = 1,$$
$$e_3' = \left(\frac{c_3}{p}\right) = \left(\frac{5}{3}\right) = \left(\frac{2}{3}\right) = -1.$$

then she gets

$$m_2 = 0, \quad \text{since } e_2' = 1,$$

$$m_3 = 1, \quad \text{since } e_3' = -1.$$

Remark 2.9. The scheme introduced above is a good extension of the public-key idea but encrypts messages bit by bit. It is completely secure with respect to semantic security as well as bit security.[1] However, a major disadvantage of the scheme is the message expansion by a factor of $\log n$ bit. To improve the efficiency of the scheme, Blum and Goldwasser [8] proposed in 1984 another randomized encryption scheme, in which the ciphertext is only longer than the plaintext by a constant number of bits; this scheme is comparable to the RSA scheme, both in terms of speed and message expansion.

Problems and Exercises for Sect. 2.2

1. The RSA function $M \mapsto C \bmod n$ is a trap-door one-way, as it is computationally intractable to invert the function if the prime factorization $n = pq$ is unknown. Give your own trap-door one-way functions that can be used to construct public-key cryptosystems. Justify your answer.

2. Show that
$$M \equiv M^{ed} \ (\text{mod } n),$$
where $ed \equiv 1 \ (\text{mod } \phi(n))$.

3. Let the ciphertexts $C_1 \equiv M_1^e \ (\text{mod } n)$ and $C_2 \equiv M_2^e \ (\text{mod } n)$ be as follows, where $e = 9137$ and n is the following RSA-129 number:

    46604906435060096392391122387112023736039163470082768_
    24341038329668507346202721798200029792506708833728356_
    7804532383891140719579,

    65064096938511069741528313342475396648978551735813836_
    77796350373814720928779386178787818974157439185718360_
    819612416009343880158.

    Find M_1 and M_2.

---

[1] Bit security is a special case of semantic security. Informally, bit security is concerned with not only that the whole message is not recoverable but also that individual bits of the message are not recoverable. The main drawback of the scheme is that the encrypted message is much longer than its original plaintext.

4. Let

$$e_1 = 9007,$$
$$e_2 = 65537,$$
$$n = 1143816257578888676692357799761466120102182967212423625625618429357069352457338978305971235639587050589890751475992900268795435 41,$$
$$C_1 \equiv M^{e_1} \pmod{n},$$
$$\equiv 104202250941196238413638382607974125774449084724929591257433745889265297771717182413024642938078351979089945343407464161377977212,$$
$$C_2 \equiv M^{e_2} \bmod n$$
$$\equiv 7645275072918870018071997051754457471094475731790989604134098748828557319028078348030908497802156339649075975060051949607130434 8.$$

Find the plaintext M.

5. (Rivest) Let
$$k = 2^{2^t} \pmod{n}$$
where
$$n = 6314466083072888893799357126131292332363298818330841375588990772701957128924885547308446055753206513618346628848948088663500368480396588171361987660521897267810162280557475393838308261759713218926668611776954526391570120690939973680089721274464666423319187806830552067951253070082020241246233982410737753705127344494169501180975241890667963858754856319805507273709904397119733614666701543905360152543373982524579313575317653646331989064651402133985265800341991903982192844710212464887459388853582070318084289023209710907032396934919962778995323320184064522476463966355937367009369212758092086293198727008292431243681,$$
$$t = 79685186856218.$$

Find k. (Note that to find k, one needs to find $2^t \pmod{\phi(n)}$ first; however, to find $\phi(n)$ one needs to factor n first.)

6. (Knuth) Let
$$\{C_1, C_2\} \equiv \{M_1^3, M_2^3\} \bmod n$$

where

$$C_1 \;=\; 6875028364370892898789953506044079907168981402585834 43$$
$$035535588237479271080090293049630566651268112334056274$$
$$332612142823187203731181519639442616568998924368271227$$
$$51237714587973722992041257530236659548756413 82171$$

$$C_2 \;=\; 7130139886169274645420466503586462247282166640137557 78$$
$$567223219797011593220849557864249703775331317377532696$$
$$534879739201868887567829519032681632688812750060251822$$
$$38844628661575836049316280566866996833345192 94663$$

$$n \;=\; 7790302288510159542362475654705578362485767620973983 94$$
$$108440222213572872511709998585048387648131944340510932$$
$$265136815168574119934775586854274094225644500087912723$$
$$2585749337061853958340278434058208881085485078737.$$

Find $\{M_1,\ M_2\}$. (Note that there are two known ways to find $\{M_1,\ M_2\}$:

$$M_i \equiv \sqrt[3]{C_i} \ (\text{mod } n),$$

$$M_i \equiv C_i^d \ (\text{mod } n),$$

where $i = 1, 2$. But in either way, one needs to find n first.

7. The original version of the RSA cryptosystem:

$$C \equiv M^e \ (\text{mod } n), \quad M \equiv C^d \ (\text{mod } n),$$

with

$$ed \equiv 1 \ (\text{mod } \phi(n))$$

is a type of deterministic cryptosystem, in which the same ciphertext is obtained for the same plaintext even at a different time. That is,

$$M_1 \xrightarrow{\text{Encryption at Time 1}} C_1 \,,$$

$$M_1 \xrightarrow{\text{Encryption at Time 2}} C_1 \,,$$

$$\vdots$$

$$M_1 \xrightarrow{\text{Encryption at Time } t} C_1 \,.$$

A randomized cryptosystem is one in which different ciphertext is obtained at a different time even for the same plaintext

$$M_1 \xrightarrow{\text{Encryption at Time 1}} C_1 \,,$$

$$M_1 \xrightarrow{\text{Encryption at Time 2}} C_2 \ ,$$

$$\vdots$$

$$M_1 \xrightarrow{\text{Encryption at Time } t} C_t \ ,$$

with $C_1 \neq C_2 \neq \cdots \neq C_t$. Describe a method to make RSA a randomized cryptosystem.

8. Describe a man-in-the-middle attack on the original version of the RSA cryptosystem.

9. Show that cracking RSA or any IFP-based cryptography is generally equivalent to solving the IFP problem.

10. Let

$n = 21290246318258757547497882016271517497806703963277216278233$
$38321538470570413250102890108976982548192582551350925260096$
$0236998394402433590752 9$

$C \equiv M^2 \pmod{n}$

$= 51285205060243481188122109876540661122140906807437327290641$
$60633920242479741450841196687149365272035106423411648279 36$
3932042884271651389234

Find the plaintext M.

2.3 Quantum Attacks on IFP and IFP-Based Cryptography

As the security of RSA or any IFP-related cryptography relies on the intractability of the IFP problem, if IFP can be solved in polynomial time, all the IFP-related cryptography can be broken efficiently in polynomial time. In this section, we discuss quantum attacks on IFP and IFP-related cryptography.

Relationships Between IFP and IFP-Based Cryptography

As can be seen, IFP is a conjectured (i.e., unproved) infeasible problem in computational number theory; this would imply that the cryptographic system-based DLP is secure and unbreakable in polynomial time:

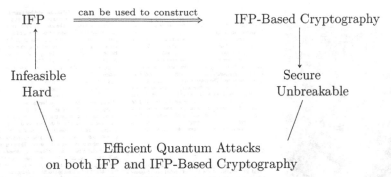

Thus, anyone who can solve IFP can break IFP-based cryptography. With this regard, solving IFP is equivalent to breaking IFP-based cryptography. As everybody knows at present, no efficient algorithm is known for solving IFP, therefore, no efficient algorithm for cracking IFP-based cryptography. However, Shor [73] showed that IFP can be solved in $\mathcal{BQP}$, where $\mathcal{BQP}$ is the class of problem that is efficiently solvable in polynomial time on a quantum Turing machine (see Fig. 2.8).

Hence, all IFP-based cryptographic systems can be broken in polynomial time on a quantum computer. Incidentally, the quantum factoring attack is intimately connected to the order finding problem which can be done in polynomial time on a quantum computer. More specifically, using the quantum order finding algorithm, the quantum factoring attack can break all IFP-based cryptographic systems, such as RSA and Rabin, which can be broken completely in polynomial time on a quantum computer :

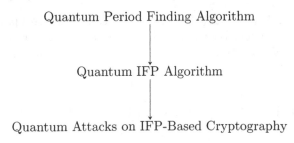

Algorithms for Quantum Computation:
Discrete Logarithms and Factoring

Peter W. Shor
AT&T Bell Labs
Room 2D-149
600 Mountain Ave.
Murray Hill, NJ 07974, USA

Abstract

A computer is generally considered to be a universal computational device; i.e., it is believed able to simulate any physical computational device with a cost in computation time of at most a polynomial factor. It is not clear whether this is still true when quantum mechanics is taken into consideration. Several researchers, starting with David Deutsch, have developed models for quantum mechanical computers and have investigated their computational properties. This paper gives Las Vegas algorithms for finding discrete logarithms and factoring integers on a quantum computer that take a number of steps which is polynomial in the input size, e.g., the number of digits of the integer to be factored. These two problems are generally considered hard on a classical computer and have been used as the basis of several proposed cryptosystems. (We thus give the first examples of quantum cryptanalysis.)

1 Introduction

Since the discovery of quantum mechanics, people have found the behavior of the laws of probability in quantum mechanics counterintuitive. Because of this behavior, quantum mechanical phenomena behave quite differently than the phenomena of classical physics that we are used to. Feynman seems to have been the first to ask what effect this has on computation [13, 14]. He gave arguments as to why this behavior might make it intrinsically computationally expensive to simulate quantum mechanics on a classical (or von Neumann) computer. He also suggested the possibility of using a computer based on quantum mechanical principles to avoid this problem, thus implicitly asking the converse question: by using quantum mechanics in a computer can you compute more efficiently than on a classical computer. Other early work in the field of quantum mechanics and computing was done by Benioff

[1, 2]. Although he did not ask whether quantum mechanics conferred extra power to computation, he did show that a Turing machine could be simulated by the reversible unitary evolution of a quantum process, which is a necessary prerequisite for quantum computation. Deutsch [9, 10] was the first to give an explicit model of quantum computation. He defined both quantum Turing machines and quantum circuits and investigated some of their properties.

The next part of this paper discusses how quantum computation relates to classical complexity classes. We will thus first give a brief intuitive discussion of complexity classes for those readers who do not have this background. There are generally two resources which limit the ability of computers to solve large problems: time and space (i.e., memory). The field of analysis of algorithms considers the asymptotic demands that algorithms make for these resources as a function of the problem size. Theoretical computer scientists generally classify algorithms as efficient when the number of steps of the algorithms grows as a polynomial in the size of the input. The class of problems which can be solved by efficient algorithms is known as P. This classification has several nice properties. For one thing, it does a reasonable job of reflecting the performance of algorithms in practice (although an algorithm whose running time is the tenth power of the input size, say, is not truly efficient). For another, this classification is nice theoretically, as different reasonable machine models produce the same class P. We will see this behavior reappear in quantum computation, where different models for quantum machines will vary in running times by no more than polynomial factors.

There are also other computational complexity classes discussed in this paper. One of these is PSPACE, which are those problems which can be solved with an amount of memory polynomial in the input size. Another important complexity class is NP, which intuitively is the class of exponential search problems. These are problems which may require the search of an exponential size space to find

0272-5428/94 $04.00 © 1994 IEEE

124

Figure 2.8. David Deutsch and the first page of his 1985 paper

Order Finding Problem

We first present some basic concept of the *order* of an element in a multiplicative group.

Definition 2.4. Let $G = \mathbb{Z}_N^*$ be a finite multiplicative group, and $x \in G$ a randomly chosen integer (element). Then order of x in G, or order of an element a modulo N, sometimes denoted by $\operatorname{order}(x, N)$, is the smallest positive integer r such that

$$x^r \equiv 1 \pmod{N}.$$

Example 2.9. Let $5 \in \mathbb{Z}_{104}^*$. Then $\operatorname{order}(5, 104) = 4$, since 4 is the smallest positive integer satisfying

$$5^4 \equiv 1 \pmod{104}.$$

Theorem 2.7. Let G be a finite group and suppose that $x \in G$ has finite order r. If $x^k = 1$, then $r \mid k$.

Example 2.10. Let $5 \in \mathbb{Z}_{104}^*$. As $5^{24} \equiv 1 \pmod{104}$, so, $4 \mid 24$.

Definition 2.5. Let G be a finite group, then the number of elements in G, denoted by $|G|$, is called the *order* of G.

Example 2.11. Let $G = \mathbb{Z}_{104}^*$. Then there are 48 elements in G that are relatively prime to 104 (two numbers a and b are relatively prime if $\gcd(a, b) = 1$), namely,

1, 3, 5, 7, 9, 11, 15, 17, 19, 21, 23, 25, 27, 29, 31, 33, 35, 37, 41, 43
45, 47, 49, 51, 53, 55, 57, 59, 61, 63, 67, 69, 71, 73, 75, 77, 79, 81
83, 85, 87, 89, 93, 95, 97, 99, 101, 103

Thus, $|G| = 48$. That is, the order of the group G is 48.

Theorem 2.8 (Lagrange). Let G be a finite group. Then the order of an element $x \in G$ divides the order of the group G.

Example 2.12. Let $G = \mathbb{Z}_{104}^*$. Then the order of G is 48, whereas the order of the element $5 \in G$ is 4. Clearly $4 \mid 24$.

Corollary 2.1. If a finite group G has order r, then $x^r = 1$ for all $x \in G$.

Example 2.13. Let $G = \mathbb{Z}_{104}^*$ and $|G| = 48$. Then

$$1^{48} \equiv 1 \pmod{104}$$
$$3^{48} \equiv 1 \pmod{104}$$
$$5^{48} \equiv 1 \pmod{104}$$
$$7^{48} \equiv 1 \pmod{104}$$

$$\vdots$$

$$101^{48} \equiv 1 \pmod{104}$$
$$103^{48} \equiv 1 \pmod{104}.$$

Now, we are in a position to present our two main theorems as follows.

Theorem 2.9. Let C be the RSA ciphertext, and $\mathrm{order}(C, N)$ the order of $C \in \mathbb{Z}_N^*$. Then

$$d \equiv 1/e \pmod{\mathrm{order}(C, N)}.$$

Corollary 2.2. Let C be the RSA ciphertext, and $\mathrm{order}(C, N)$ the order of $C \in \mathbb{Z}_N^*$. Then

$$M \equiv C^{1/e \ (\mathrm{mod} \ \mathrm{order}(C,N))} \pmod{N}$$

Thus, to recover the RSA M from C, it suffices to just find the order of C modulo N.

Now, we return to the actual computation of the order of an element x in $G = \mathbb{Z}_N^*$. Finding the order of an element x in G is, in theory, not a problem: Just keep multiplying until we get to "1," the identity element of the multiplicative group G. For example, let $N = 179359$, $x = 3 \in G$, and $G = \mathbb{Z}^*_{179359}$, such that $\gcd(3, 179359) = 1$. To find the order $r = \text{order}(3, 179359)$, we just keep multiplying until we get to "1":

$$
\begin{array}{rclcl}
3^1 & \text{mod} & 179359 & = & 3 \\
3^2 & \text{mod} & 179359 & = & 9 \\
3^3 & \text{mod} & 179359 & = & 27 \\
& & \vdots & & \\
3^{1000} & \text{mod} & 179359 & = & 31981 \\
3^{1001} & \text{mod} & 179359 & = & 95943 \\
3^{1002} & \text{mod} & 179359 & = & 108470 \\
& & \vdots & & \\
3^{14716} & \text{mod} & 179359 & = & 99644 \\
3^{14717} & \text{mod} & 179359 & = & 119573 \\
3^{14718} & \text{mod} & 179359 & = & 1.
\end{array}
$$

Thus, the order r of 3 in the multiplicative group $G = (\mathbb{Z}/179359\mathbb{Z})^*$ is 14718, that is, $\text{ord}_{179359}(3) = 14718$.

Example 2.14. Let

$$
\begin{aligned}
N &= 5515596313 \\
e &= 1757316971 \\
C &= 763222127 \\
r &= \text{order}(C, N) = 114905160
\end{aligned}
$$

Then

$$
\begin{aligned}
M &\equiv C^{1/e \bmod r} \pmod{N} \\
&\equiv 763222127^{1/1757316971 \bmod 114905160} \pmod{5515596313} \\
&\equiv 1612050119
\end{aligned}
$$

Clearly, this result is correct, since

$$
\begin{aligned}
M^e &\equiv 1612050119^{1757316971} \\
&\equiv 763222127 \\
&\equiv C \pmod{5515596313}
\end{aligned}
$$

It must also be noted, however, that in practice, the above computation for finding the order of $x \in (\mathbb{Z}/N\mathbb{Z})^*$ may not work, since for an element x in a large group G with N having more than 200 digits, the computation of r may require more than 10^{150} multiplications. Even if these multiplications could be carried out at the rate of $1,000$ billion/s on a supercomputer, it would take approximately $3 \cdot 10^{80}$ years to arrive at the answer. Thus, the order finding problem is intractable on conventional digital computers. The problem is, however, tractable on quantum computers, provided that a practical quantum computer is available.

It is worthwhile pointing out that although the order is hard to find, the exponentiation is easy to compute. Suppose we want to compute $x^e \bmod n$ with $x, e, n \in \mathbb{N}$. Suppose moreover that the binary form of e is as follows:

$$e = \beta_k 2^k + \beta_{k-1} 2^{k-1} + \cdots + \beta_1 2^1 + \beta_0 2^0, \tag{2.48}$$

where each β_i $(i = 0, 1, 2, \cdots k)$ is either 0 or 1. Then we have

$$
\begin{aligned}
x^e &= x^{\beta_k 2^k + \beta_{k-1} 2^{k-1} + \cdots + \beta_1 2^1 + \beta_0 2^0} \\
&= \prod_{i=0}^{k} x^{\beta_i 2^i} \\
&= \prod_{i=0}^{k} \left(x^{2^i} \right)^{\beta_i}.
\end{aligned} \tag{2.49}
$$

Furthermore, by the exponentiation law,

$$x^{2^{i+1}} = \left(x^{2^i} \right)^2, \tag{2.50}$$

and so the final value of the exponentiation can be obtained by *repeated squaring and multiplication* operations. For example, to compute a^{100}, we first write $100_{10} = 1100100_2 := e_6 e_5 e_4 e_3 e_2 e_1 e_0$, and then compute

$$a^{100} = \left(\left(\left(\left(\left((a)^2 \cdot a \right)^2 \right)^2 \right)^2 \cdot a \right)^2 \right)^2 \tag{2.51}$$

$$\Rightarrow \quad a, \ a^3, \ a^6, \ a^{12}, \ a^{24}, \ a^{25}, \ a^{50}, \ a^{100}.$$

Note that for each e_i, if $e_i = 1$, we perform a *squaring* and a *multiplication* operation (except "$e_6 = 1$," for which we just write down a, as indicated in the first bracket); otherwise, we perform only a *squaring* operation. That is,

e_6	1	a	a	initialization
e_5	1	$(a)^2 \cdot a$	a^3	squaring and multiplication
e_4	0	$\left((a)^2 \cdot a \right)^2$	a^6	squaring
e_3	0	$\left(\left((a)^2 \cdot a \right)^2 \right)^2$	a^{12}	squaring
e_2	1	$\left(\left(\left((a)^2 \cdot a \right)^2 \right)^2 \right)^2 \cdot a$	a^{25}	squaring and multiplication
e_1	0	$\left(\left(\left(\left((a)^2 \cdot a \right)^2 \right)^2 \right)^2 \cdot a \right)^2$	a^{50}	squaring
e_0	0	$\left(\left(\left(\left(\left((a)^2 \cdot a \right)^2 \right)^2 \right)^2 \cdot a \right)^2 \right)^2$	a^{100}	squaring

$$\shortparallel$$
$$a^{100}$$

The following is the algorithm, which runs in in $\mathcal{O}(\log e)$ arithmetic operations and $\mathcal{O}\left((\log e)(\log n)^2 \right)$ bit operations.

Algorithm 2.7 (Fast modular exponentiation $x^e \bmod n$). This algorithm will compute the modular exponentiation

$$c \equiv x^e \pmod{n},$$

where $x, e, n \in \mathbb{N}$ with $n > 1$. It requires at most $2 \log e$ and $2 \log e$ divisions (divisions are only needed for modular operations; they can be removed if only $c = x^e$ are required to be computed).

[1] [Precomputation] Let

$$e_{\beta-1} e_{\beta-2} \cdots e_1 e_0 \tag{2.52}$$

be the binary representation of e (i.e., e has β bits). For example, for $562 = 1000110010$, we have $\beta = 10$ and

1	0	0	0	1	1	0	0	1	0
↑	↑	↑	↑	↑	↑	↑	↑	↑	↑
e_9	e_8	e_7	e_6	e_5	e_4	e_3	e_2	e_1	e_0

[2] [Initialization] Set $c \leftarrow 1$.

[3] [Modular Exponentiation] Compute $c = x^e \bmod n$ in the following way:

> for i from $\beta - 1$ down to 0 do
> $\quad c \leftarrow c^2 \bmod n$ (squaring)
> $\quad$ if $e_i = 1$ then
> $\qquad c \leftarrow c \cdot x \bmod n$ (multiplication)

[4] [Exit] Print c and terminate the algorithm.

Quantum Order Computing

It may be the case that, as the famous physicist Feynman mentioned, nobody understands quantum mechanics, some progress has been made in quantum mechanics, particularly in quantum computing and quantum cryptography. In this section, we present a quantum algorithm for computing the order of an element x in the multiplicative group $\mathbb{Z}_N^*$, due to Shor [69]. The main idea of Shor's algorithm is as follows. First of all, we create two quantum registers for our quantum computer: Register-1 and Register-2. Of course, we can create just one single quantum memory register partitioned into two parts. Secondly, we create in Register-1 a superposition of the integers $a = 0, 1, 2, 3, \cdots$ which will be the arguments of $f(a) = x^a \pmod{N}$, and load Register-2 with all zeros. Thirdly, we compute in Register-2 $f(a) = x^a \pmod{N}$ for each input a. (Since the values of a are kept in Register-1, this can be done reversibly.) Fourthly, we perform the discrete Fourier transform on Register-1. Finally, we observe both registers of the machine and find the order r that satisfies $x^r \equiv 1 \pmod{N}$. The following is a brief description of the quantum algorithm for the order finding problem.

Algorithm 2.8 (Quantum order finding attack). Given RSA ciphertext C and modulus N, this attack will first find the order, r, of C in $\mathbb{Z}_N^8$, such that $C^r \equiv 1 \pmod{N}$, then recover the plaintext M from the ciphertext C. Assume the quantum computer has two quantum registers: Register-1 and Register-2, which hold integers in binary form.

[1] (Initialization) Find a number q, a power of 2, say 2^t, with $N^2 < q < 2N^2$.

[2] (Preparation for quantum registers) Put in the first t-qubit register, Register-1, the uniform superposition of states representing numbers $a \pmod{q}$, and load Register-2 with all zeros. This leaves the machine in the state $|\Psi_1\rangle$:

$$|\Psi_1\rangle = \frac{1}{\sqrt{q}} \sum_{a=0}^{q-1} |a\rangle |0\rangle.$$

(Note that the joint state of both registers are represented by $|\text{Register-1}\rangle$ and $|\text{Register-2}\rangle$). What this step does is put each qubit in Register-1 into the superposition

$$\frac{1}{\sqrt{2}} (|0\rangle + |1\rangle).$$

[3] (Power Creation) Fill in the second t-qubit register, Register-2, with powers $C^a \pmod{N}$. This leaves the machine in state $|\Psi_2\rangle$:

$$|\Psi_2\rangle = \frac{1}{\sqrt{q}} \sum_{a=0}^{q-1} |a\rangle |C^a \pmod{N}\rangle.$$

This step can be done reversibly since all the a's were kept in Register-1.

[4] (Perform a quantum FFT) Apply FFT on Register-1. The FFT maps each state $|a\rangle$ to

$$\frac{1}{\sqrt{q}} \sum_{c=0}^{q-1} \exp(2\pi i a c / q) |c\rangle.$$

That is, we apply the unitary matrix with the (a, c) entry equal to $\frac{1}{\sqrt{q}} \exp(2\pi i a c / q)$. This leaves the machine in the state $|\Psi_3\rangle$:

$$|\Psi_3\rangle = \frac{1}{q} \sum_{a=0}^{q-1} \sum_{c=0}^{q-1} \exp(2\pi i a c / q) |c\rangle |C^a \pmod{N}\rangle.$$

[5] (Periodicity Detection in x^a) Observe both $|c\rangle$ in Register-1 and $|C^a \pmod{N}\rangle$ in Register-2 of the machine, measure both arguments of this superposition, obtaining the values of $|c\rangle$ in the first argument and some $|x^k \pmod{n}\rangle$ as the answer for the second one $(0 < k < r)$.

[6] (Extract r) Extract the required value of r. Given the pure state $|\Psi_3\rangle$, the probabilities of different results for this measurement will be given by the probability distribution:

$$\text{Prob}(c, C^k \ (\text{mod } N)) = \left| \frac{1}{q} \sum_{\substack{a=0 \\ C^a \equiv a^k \ (\text{mod } N)}}^{q-1} \exp(2\pi i a c/q) \right|^2$$

$$= \left| \frac{1}{q} \sum_{B=0}^{\lfloor (q-k-1)/r \rfloor} \exp(2\pi i(br+k)c/q) \right|^2$$

$$= \left| \frac{1}{q} \sum_{B=0}^{\lfloor (q-k-1)/r \rfloor} \exp(2\pi i b\{rc\}/q) \right|^2$$

where $\{rc\}$ is $rc \bmod N$. As shown in [69],

$$\frac{-r}{2} \leqslant \{rc\} \leqslant \frac{-r}{2} \implies \frac{-r}{2} \leqslant rc - dq \leqslant \frac{-r}{2}, \text{ for some } d$$

$$\implies \text{Prob}(c, C^k \ (\text{mod } N)) > \frac{1}{3r^2}.$$

then we have

$$\left| \frac{c}{q} - \frac{d}{r} \right| \leqslant \frac{1}{2q}.$$

Since $\frac{c}{q}$ were known, r can be obtained by the continued fraction expansion of $\frac{c}{q}$.

[7] (Code Breaking) Once the order r, $r = \text{order}(C, N)$, is found, then compute:

$$M \equiv C^{1/e \bmod r} \ (\text{mod } N).$$

Hence, decodes the RSA code C.

Theorem 2.10. (Complexity of Quantum Order Finding Attack). Quantum order attack can find $\text{order}(C, N)$ and recover M from C in time $\mathcal{O}((\log N)^{2+\epsilon})$.

Remark 2.10. This quantum attack is for particular RSA ciphertexts C. In this special case, the factorization of the RSA modulus N is not needed. In the next section, we shall consider the more general quantum attack by factoring N.

Quantum Integer Factorization

Instead of finding the order of C in $\mathbb{Z}_N^*$, one can take this further to a more general case: find the order of an element x in $\mathbb{Z}_N^*$, denoted by order(x, N), where N is the RSA modulus. Once the order of an element x in $\mathbb{Z}_N^*$ is found, and it is even, it will have a good chance to factor N, of course in polynomial time, by just calculating

$$\left\{ \gcd(x^{r/2} + 1, N), \quad \gcd(x^{r/2} - 1, N) \right\}.$$

For instance, for $x = 3$, $r = 14718$, and $N = 179359$, we have

$$\left\{ \gcd(3^{14718/2} + 1, 179359), \quad \gcd(3^{14718/2} - 1, 179359) \right\} = (67, 2677),$$

and hence the factorization of N:

$$N = 179359 = 67 \cdot 2677.$$

The following theorem shows that the probability for r to work is high.

Theorem 2.11. Let the odd integer $N > 1$ have exactly k distinct prime factors. For a randomly chosen $x \in \mathbb{Z}_N^*$ with multiplicative order r, the probability that r is even and that

$$x^{r/2} \not\equiv -1 \pmod{N}$$

is least $1 - 1/2^{k-1}$. More specifically, if N has just two prime factors (this is often the case for the RSA modulus N), then the probability is at least $1/2$.

Algorithm 2.9 (Quantum algorithm for integer factorization). Given integers x and N, the algorithm will

– find the order of x, i.e., the smallest positive integer r such that

$$x^r \equiv 1 \pmod{N},$$

– find the prime factors of N and compute the decryption exponent d,
– decode the RSA message.

Assume the machine has two quantum registers: Register-1 and Register-2, which hold integers in binary form.

[1] (Initialization) Find a number q, a power of 2, say 2^t, with $N^2 < q < 2N^2$.

[2] (Preparation for quantum registers) Put in the first t-qubit register, Register-1, the uniform superposition of states representing numbers $a \pmod{q}$, and load Register-2 with all zeros. This leaves the machine in the state $|\Psi_1\rangle$:

$$|\Psi_1\rangle = \frac{1}{\sqrt{q}} \sum_{a=0}^{q-1} |a\rangle |0\rangle.$$

(Note that the joint state of both registers are represented by $|\text{Register-1}\rangle$ and $|\text{Register-2}\rangle$). What this step does is put each qubit in Register-1 into the superposition

$$\frac{1}{\sqrt{2}} (|0\rangle + |1\rangle).$$

[3] (Base Selection) Choose a random $x \in [2, N-2]$ such that $\gcd(x, N) = 1$.

[4] (Power Creation) Fill in the second t-qubit register, Register-2, with powers $x^a \pmod{N}$. This leaves the machine in state $|\Psi_2\rangle$:

$$|\Psi_2\rangle = \frac{1}{\sqrt{q}} \sum_{a=0}^{q-1} |a\rangle |x^a \pmod{N}\rangle.$$

This step can be done reversibly since all the a's were kept in Register-1.

[5] (Perform a quantum FFT) Apply FFT on Register-1. The FFT maps each state $|a\rangle$ to

$$\frac{1}{\sqrt{q}} \sum_{c=0}^{q-1} \exp(2\pi i a c / q) |c\rangle.$$

That is, we apply the unitary matrix with the (a, c) entry equal to $\frac{1}{\sqrt{q}} \exp(2\pi i a c / q)$. This leaves the machine in the state $|\Psi_3\rangle$:

$$|\Psi_3\rangle = \frac{1}{q} \sum_{a=0}^{q-1} \sum_{c=0}^{q-1} \exp(2\pi i a c / q) |c\rangle |x^a \pmod{N}\rangle.$$

[6] (Periodicity Detection in x^a) Observe both $|c\rangle$ in Register-1 and $|x^a \pmod{N}\rangle$ in Register-2 of the machine, measure both arguments of this superposition, obtaining the values of $|c\rangle$ in the first argument and some $|x^k \pmod{n}\rangle$ as the answer for the second one ($0 < k < r$).

[7] (Extract r) Extract the required value of r. Given the pure state $|\Psi_3\rangle$, the probabilities of different results for this measurement will be given by the probability distribution:

$$\text{Prob}(c, x^k \pmod{N}) = \left| \frac{1}{q} \sum_{\substack{a=0 \\ x^a \equiv a^k \pmod{N}}}^{q-1} \exp(2\pi i a c / q) \right|^2$$

$$= \left| \frac{1}{q} \sum_{B=0}^{\lfloor (q-k-1)/r \rfloor} \exp(2\pi i (br+k) c / q) \right|^2$$

$$= \left| \frac{1}{q} \sum_{B=0}^{\lfloor (q-k-1)/r \rfloor} \exp(2\pi i b \{rc\} / q) \right|^2$$

where $\{rc\}$ is $rc \bmod N$. As showed in [69],

$$\frac{-r}{2} \leqslant \{rc\} \leqslant \frac{-r}{2} \implies \frac{-r}{2} \leqslant rc - dq \leqslant \frac{-r}{2}, \text{ for some } d$$

$$\implies \text{Prob}(c, x^k \ (\bmod \ N)) > \frac{1}{3r^2}.$$

then we have

$$\left| \frac{c}{q} - \frac{d}{r} \right| \leqslant \frac{1}{2q}.$$

Since $\frac{c}{q}$ were known, r can be obtained by the continued fraction expansion of $\frac{c}{q}$.

[8] (Resolution) If r is odd, go to Step [3] to start a new base. If r is even, then try to compute Once r is found, the factors of N can be *possibly*

$$\{\gcd(x^{r/2} - 1, N), \ \gcd(x^{r/2} + 1, N)\}$$

Hopefully, this will produce two factors of N.

[9] (Computing d) Once N is factored and p and q are found, then compute

$$d \equiv 1/e \ (\bmod \ (p-1)(q-1)).$$

[10] (Code Break) As soon as d is found, and RSA ciphertext encrypted by the public-key (e, N), can be decrypted by this d as follows:

$$M \equiv C^d \ (\bmod \ N).$$

Theorem 2.12 (Complexity of Quantum Factoring). Quantum factoring algorithm can factor the RSA modulus N and break the RSA system in time $\mathcal{O}((\log N)^{2+\epsilon})$.

Remark 2.11. The attack discussed in Algorithm 2.9 is more general than that in Algorithm 2.8. Algorithm 2.9 also implies that if a practical quantum computer can be built, then the RSA cryptosystem can be completely broken, and a quantum resistant cryptosystem must be developed and used to replace the RSA cryptosystem.

Example 2.15. On 19 December 2001, IBM made the first demonstration of the quantum factoring algorithm [77] that correctly identified 3 and 5 as the factors of 15. Although the answer may appear to be trivial, it may have good potential and practical implication. In this example, we show how to factor 15 quantum-mechanically [56]:

[1] Choose at random $x = 7$ such that $\gcd(x, N) = 1$. We aim to find $r = \text{order}_{15} 7$ such that $7^r \equiv 1 \ (\bmod \ 15)$.

[2] Initialize two four-qubit registers to state 0:

$$|\Psi_0\rangle = |0\rangle|0\rangle.$$

[3] Randomize the first register as follows:

$$|\Psi_0\rangle \rightarrow |\Psi_1\rangle = \frac{1}{\sqrt{2^t}} \sum_{k=0}^{2^t-1} |k\rangle|0\rangle.$$

[4] Unitarily compute the function $f(a) \equiv 13^a \pmod{15}$ as follows:

$$
\begin{aligned}
|\Psi_1\rangle \rightarrow |\Psi_2\rangle &= \frac{1}{\sqrt{2^t}} \sum_{k=0}^{2^t-1} |k\rangle \left|13^k \pmod{15}\right\rangle \\
&= \frac{1}{\sqrt{2^t}} \Big[\, |0\rangle|1\rangle + |1\rangle|7\rangle + |2\rangle|4\rangle + |3\rangle|13\rangle + \\
&\qquad |4\rangle|1\rangle + |5\rangle|7\rangle + |6\rangle|4\rangle + |7\rangle|13\rangle + \\
&\qquad |8\rangle|1\rangle + |9\rangle|7\rangle + |10\rangle|4\rangle + |11\rangle|13\rangle + \\
&\qquad + \cdots \Big]
\end{aligned}
$$

[5] We now apply the FFT to the second register and measure it (it can be done in the first), obtaining a random result from $1, 7, 4, 13$. Suppose we incidently get 4, then the state input to FFT would be

$$\sqrt{\frac{4}{2^t}} \, \big[\, |2\rangle + |6\rangle + |10\rangle + |14\rangle + \cdots \big].$$

After applying FFT, some state

$$\sum_\lambda \alpha_\lambda |\lambda\rangle$$

with the probability distribution for $q = 2^t = 2048$ (see [56]). The final measurement gives $0, 512, 1024, 2048$, each with probability almost exactly $1/4$. Suppose $\lambda = 1536$ was obtained from the measurement. Then we compute the continued fraction expansion

$$\frac{\lambda}{q} = \frac{1536}{2048} = \frac{1}{1 + \frac{1}{3}}, \quad \text{with convergents} \ \left[0, 1, \frac{3}{4}, \right]$$

Thus, $r = 4 = \mathrm{order}_{15}(7)$. Therefore,

$$\gcd(x^{r/2} \pm 1, N) = \gcd(7^2 \pm 1, 15) = (5, 3).$$

Remark 2.12. Quantum factoring is still in its very earlier stage and will not threaten the security of RSA at least at present, as the current quantum computer can only factor a number with only 2 digits such as 15 which is essentially hopeless.

Exercises and Problems for Sect. 2.3

1. Show that if in Shor's factoring algorithm, we have

$$\left| \frac{c}{2^m} - \frac{d}{r} \right| < \frac{1}{2n^2}$$

and

$$\left| \frac{c}{2^m} - \frac{d_1}{r_1} \right| < \frac{1}{2n^2},$$

then

$$\frac{d}{r} = \frac{d_1}{r_1}.$$

2. Show that in case $r \nmid 2^n$, Shor's factoring algorithm [70] needs to be repeated only $\mathcal{O}(\log \log r)$ steps in order to achieve the high probability of success.

3. Let $0 < s \leqslant m$. Fix an integer x_0 with $0 \leqslant x_0 < 2^s$. Show that

$$\sum_{\substack{0 \leqslant c < 2^m \\ c \equiv c_0 \ (\mathrm{mod}\ 2^s)}} e^{2\pi i c x / 2^m} = \begin{cases} 0 & \text{if } x \not\equiv 0 \ (\mathrm{mod}\ 2^{m-s}) \\ 2^{m-s} e^{2\pi i x c_0 / 2^m} & \text{if } x \equiv 0 \ (\mathrm{mod}\ 2^{m-s}) \end{cases}$$

4. There are currently many pseudo-simulations of Shor's quantum factoring algorithm; for example, the paper by Schneiderman, Stanley, and Aravind [66] gives one of the simulations in Maple, whereas Browne [12] presents an efficient classical simulation of the quantum Fourier transform based on [66]. Construct your own Java (C/C++, Mathematica or Maple) program to simulate Shor's quantum factoring algorithm and discrete logarithm algorithm.

5. Both ECM factoring algorithm and NFS factoring algorithm are very well suited for parallel implementation. Is it possible to utilize the quantum parallelism to implement ECM and NSF algorithms? If so, give a complete description the quantum ECM and NFS algorithms.

6. Pollard [58] and Strassen [75] showed that FFT can be utilized to factor an integer n in $\mathcal{O}(n^{1/4+\epsilon})$ steps, deterministically. Is it possible to replace the classical FFT with a quantum FFT in the Pollard–Strassen method, in order to obtain a deterministic quantum polynomial-time factoring algorithm (i.e., to obtain a $\mathcal{QP}$ factoring algorithm rather than the $\mathcal{BQP}$ algorithm as proposed by Shor)? If so, give a full description of the $\mathcal{QP}$ factoring algorithm.

7. At the very heart of the Pollard ρ-method for IFP lives the phenomenon of periodicity. Develop a quantum period-finding algorithm, if possible, for the ρ factoring algorithm.

2.4 Conclusions, Notes, and Further Reading

The theory of prime numbers is one of the oldest subject in number theory and indeed in the whole of mathematics, whereas the IFP is one of the oldest number-theoretic problems in the field. The root of the problem can be traced back to Euclid's *Elements* [25], although it was first clearly stated in Gauss' *Disquisitiones* [29]. With the advent of modern public-key cryptography, it has an important application in the construction of unbreakable public-key cryptographic schemes and protocols, such as RSA [28, 64], Rabin [62], and zero-knowledge proofs [33]. IFP is currently a very hot and applicable research topic, and there are many good references in the field; for a general reading, the following references are highly recommended: [1, 4, 11, 13, 17, 19, 21, 23, 40, 45, 50, 53, 61, 63, 87].

IFP-based cryptography forms an important class of public-key cryptography. In particular, RSA cryptography is the most famous and widely used cryptographic schemes in today's Internet world. More information on IFP-based cryptography can be found in [9, 20, 30, 31, 36, 37, 39, 42, 52, 76, 84], and [86].

Shor's discovery of the quantum factoring algorithm [69, 70, 70–73] in 1994 generated a great deal of research and interest in the field. Quantum computers provided a completely new paradigm for the theory of computation, and it was the first time to show that IFP can be solved efficiently in polynomial time on a quantum computer. Now, there are many good references on quantum computation, particularly on quantum factoring. Readers who wish to know more about quantum computers and quantum computation are suggested to consult the following references: [2, 5–7, 16, 22, 24, 35, 43, 48, 51, 56, 74, 77–83, 85, 88, 89], and [90]. Feynman is perhaps the father of quantum computation whose original idea about quantum computers may be found in [26, 27].

In addition to quantum computation for factoring, there are also some other non-classical computations for factoring such as molecular DNA-based factoring and attacking. For example, Chang et al. proposed some fast parallel molecular DNA algorithms for factoring large integers [14] and for breaking RSA cryptography [15].

REFERENCES

[1]   L.M. Adleman, Algorithmic number theory – the complexity contribution, in *Proceedings of the 35th Annual IEEE Symposium on Foundations of Computer Science* (IEEE, New York, 1994), pp. 88–113

[2] L.M. Adleman, J. DeMarrais, M.D.A. Huang, Quantum computability. SIAM J. Comput. **26**(5), 1524–1540 (1997)

[3] M. Agrawal, N. Kayal, N. Saxena, Primes is in P. Ann. Math. **160**(2), 781–793 (2004)

[4] D. Atkins, M. Graff, A.K. Lenstra, P.C. Leyland, The magic words are Squeamish Ossifrage, in *Advances in Cryptology – ASIACRYPT'94*. Lecture Notes in Computer Science, vol. 917 (Springer, Berlin, 1995), pp. 261–277

[5] C.H. Bennett, D.P. DiVincenzo, Quantum information and computation. Nature **404**, 247–255 (2000)

[6] C.H. Bennett, E. Bernstein et al., Strengths and weakness of quantum computing. SIAM J. Comput. **26**(5), 1510–1523 (1997)

[7] E. Bernstein, U. Vazirani, Quantum complexity theory. SIAM J. Comput. **26**(5), 1411–1473 (1997)

[8] M. Blum, S. Goldwasser, An efficient probabilistic public-key encryption scheme that hides all partial information, in *Advances in Cryptography, CRYPTO '84*. Proceedings, Lecture Notes in Computer Science, vol. 196 (Springer, Berlin, 1985), pp. 289–302

[9] D. Boneh, Twenty years of attacks on the RSA cryptosystem. Not. AMS **46**(2), 203–213 (1999)

[10] R.P. Brent, An improved Monte Carlo factorization algorithm. BIT **20**, 176–184 (1980)

[11] D.M. Bressound, *Factorization and Primality Testing* (Springer, New York, 1989)

[12] D.E. Browne, Efficient classical simulation of the quantum Fourier transform. New J. Phys. **9**, 146, 1–7 (2007)

[13] J.P. Buhler, P. Stevenhagen (eds.), *Algorithmic Number Theory* (Cambridge University Press, Cambridge, 2008)

[14] W.L. Chang, M. Guo, M.S.H. Ho, Fast parallel molecular algorithms for DNA-based computation: factoring integers. IEEE Trans. Nanobioscience 4(2), 149–163 (2005)

[15] W.L. Chang, K.W. Lin et al., Molecular solutions of the RSA public-key cryptosystem on a DNA-based computer. J. Supercomput. **56**(2), 129–163 (2011)

[16] I.L. Chuang, R. Laflamme, P. Shor, W.H. Zurek, Quantum computers, factoring, and decoherence. Science **270**, 1633–1635 (1995)

[17] H. Cohen, in *A Course in Computational Algebraic Number Theory*. Graduate Texts in Mathematics, vol. 138 (Springer, Berlin, 1993)

[18] D. Coppersmith, Small solutions to polynomial equations, and low exponent RSA vulnerability. J. Cryptol. **10**, 233–260 (1997)

[19] T.H. Cormen, C.E. Ceiserson, R.L. Rivest, *Introduction to Algorithms*, 3rd edn. (MIT, Cambridge, 2009)

[20] J.S. Coron, A. May, Deterministic polynomial-time equivalence of computing the RSA secret key and factoring. J. Cryptol. **20**(1), 39–50 (2007)

[21] R. Crandall, C. Pomerance, *Prime Numbers – A Computational Perspective*, 2nd edn. (Springer, Berlin, 2005)

[22] D. Deutsch, Quantum theory, the Church–Turing principle and the universal quantum computer. Proc. R. Soc. Lond. Ser. A **400**, 96–117 (1985)

[23] J.D. Dixon, Factorization and primality tests. Am. Math. Mon. **91**(6), 333–352 (1984)

[24] A. Ekert, R. Jozsa, Quantum computation and Shor's factoring algorithm. SIAM J. Comput. **26**(5), 1510–1523 (1997)

[25] Euclid, in *The Thirteen Books of Euclid's Elements*, 2nd edn. Translated by T.L. Heath. Great Books of the Western World, vol. 11 (William Benton Publishers, New York, 1952)

[26] R.P. Feynman, Simulating physics with computers. Int. J. Theor. Phys. **21**, 467–488 (1982)

[27] R.P. Feynman, in *Feynman Lectures on Computation*, ed. by A.J.G. Hey, R.W. Allen (Addison-Wesley, Reading, 1996)

[28] M. Gardner, Mathematical games – a new kind of Cipher that would take millions of years to break. Sci. Am. **237**(2), 120–124 (1977)

[29] C.F. Gauss, *Disquisitiones Arithmeticae*, G. Fleischer, Leipzig, 1801. English translation by A.A. Clarke (Yale University Press, Yale, 1966) Revised English translation by W.C. Waterhouse (Springer, Berlin, 1975)

[30] O. Goldreich, *Foundations of Cryptography: Basic Tools* (Cambridge University Press, Cambridge, 2001)

[31] O. Goldreich, *Foundations of Cryptography: Basic Applications* (Cambridge University Press, Cambridge, 2004)

[32] S. Goldwasser, S. Micali, Probabilistic encryption. J. Comput. Syst. Sci. **28**, 270–299 (1984)

[33] S. Goldwasser, S. Micali, C. Rackoff, The knowledge complexity of interactive proof systems. SIAM J. Comput. **18**(1), 186–208 (1989)

[34] J. Grobchadl, The Chinese remainder theorem and its application in a high-speed RSA Crypto chip, in *Proceedings of the 16th Annual Computer Security Applications Conference (ACSAC'00)* (IEEE, New York, 2000), pp. 384–393

[35] J. Grustka, *Quantum Computing* (McGraw-Hill, New York, 1999)

[36] M.J. Hinek, *Cryptanalysis of RSA and Its Variants* (Chapman & Hall/CRC Press, London/West Palm Beach, 2009)

[37] J. Hoffstein, J. Pipher, J.H. Silverman, *An Introduction to Mathematical Cryptography* (Springer, Berlin, 2008)

[38] K. Ireland, M. Rosen, in *A Classical Introduction to Modern Number Theory*, 2nd edn. Graduate Texts in Mathematics, vol. 84 (Springer, Berlin, 1990)

[39] S. Katzenbeisser, *Recent Advances in RSA Cryptography* (Kluwer, Dordrecht, 2001)

[40] T. Kleinjung et al., Factorization of a 768-bit RSA modulus, in *CRYPTO 2010*, ed. by T. Rabin. Lecture Notes in Computer Science, vol. 6223 (Springer, New York, 2010), pp. 333–350

[41] D.E. Knuth, *The Art of Computer Programming III – Sorting and Searching*, 2nd edn. (Addison-Wesley, Reading, 1998)

[42] A.G. Konheim, *Computer Security and Cryptography* (Wiley, New York, 2007)

[43] B.P. Lanyou, T.J. Weinhold et al., Experiemntal demonstration of a compiled version of Shor's algorithm' with quantum entabglement. Phys. Rev. Lett. **99**, 250504, 4 (2007)

[44] R.S. Lehman, Factoring large integers. Math. Comput. **28**, 126, 637–646 (1974)

[45] A.K. Lenstra, Integer factoring. Des. Codes Cryptography **19**(2/3), 101–128 (2000)

[46] A.K. Lenstra, H.W. Lenstra Jr. (eds.), in *The Development of the Number Field Sieve*. Lecture Notes in Mathematics, vol. 1554 (Springer, Berlin, 1993)

[47] H.W. Lenstra Jr., Factoring integers with elliptic curves. Ann. Math. **126**, 649–673 (1987)

[48] S.J. Lomonaco Jr., Shor's quantum factoring algorithm. AMS Proc. Symp. Appl. Math. **58**, 19 (2002)

[49] J.F. McKee, Turning Euler's factoring methods into a factoring algorithm. Bull. Lond. Math. Soc. **28**, 351–355 (1996)

[50] J.F. McKee, R. Pinch, Old and new deterministic factoring algorithms, in *Algorithmic Number Theory*. Lecture Notes in Computer Science, vol. 1122 (Springer, Berlin, 1996), pp. 217–224

[51] N.D. Mermin, *Quantum Computer Science* (Cambridge University Press, Cambridge, 2007)

[52] R.A. Mollin, *RSA and Public-Key Cryptography* (Chapman & Hall/CRC Press, London/West Palm Beach, 2003)

[53] P.L. Montgomery, Speeding Pollard's and elliptic curve methods of factorization. Math. Comput. **48**, 243–264 (1987)

[54] P.L. Montgomery, A survey of modern integer factorization algorithms. CWI Q. **7**(4), 337–394 (1994)

[55] M.A. Morrison, J. Brillhart, A method of factoring and the factorization of F_7. Math. Comput. **29**, 183–205 (1975)

[56] M.A. Nielson, I.L. Chuang, *Quantum Computation and Quantum Information*, 10th Anniversary edn. (Cambridge University Press, Cambridge, 2010)

[57] S.C. Pohlig, M. Hellman, An improved algorithm for computing logarithms over GF(p) and its cryptographic significance. IEEE Trans. Inf. Theor. **24**, 106–110 (1978)

[58] J.M. Pollard, Theorems on factorization and primality testing. Proc. Camb. Phil. Soc. **76**, 521–528 (1974)

[59] J.M. Pollard, A Monte Carlo method for factorization. BIT **15**, 331–332 (1975)

[60] C. Pomerance, The quadratic Sieve factoring algorithm, in *Proceedings of Eurocrypt 84*. Lecture Notes in Computer Science, vol. 209 (Springer, Berlin, 1985), pp. 169–182

[61] C. Pomerance, A tale of two sieves. Not. AMS **43**(12), 1473–1485 (1996)

[62] M. Rabin, Digitalized Signatures and Public-Key Functions as Intractable as Factorization. Technical Report MIT/LCS/TR-212, MIT Laboratory for Computer Science (1979)

[63] H. Riesel, *Prime Numbers and Computer Methods for Factorization* (Birkhäuser, Boston, 1990)

[64] R.L. Rivest, A. Shamir, L. Adleman, A method for obtaining digital signatures and public key cryptosystems. Comm. ACM **21**(2), 120–126 (1978)

[65] R.L. Rivest, B. Kaliski, RSA Problem, in *Encyclopedia of Cryptography and Security*, ed. by H.C.A. van Tilborg (Springer, Berlin, 2005)

[66] J.F. Schneiderman, M.E. Stanley, P.K. Aravind, A pseudo-simulation of Shor's quantum factoring algorithm, 20 pages (2002) [arXiv:quant-ph/0206101v1]

[67] D. Shanks, class number, a theory of factorization, and genera, in *Proceedings of Symposium of Pure Mathematics*, vol. XX, State Univ. New York, Stony Brook, 1969 (American Mathematical Society, Providence, 1971), pp. 415–440

[68] D. Shanks, Analysis and improvement of the continued fraction method of factorization, Abstract 720-10-43. Am. Math. Soc. Not. **22**, A-68 (1975)

[69] P. Shor, Algorithms for quantum computation: discrete logarithms and factoring, in *Proceedings of 35th Annual Symposium on Foundations of Computer Science* (IEEE Computer Society, Silver Spring, 1994), pp. 124–134

[70] P. Shor, Polynomial-time algorithms for prime factorization and discrete logarithms on a quantum computer. SIAM J. Comput. **26**(5), 1484–1509 (1997)

[71] P. Shor, Quantum computing. Documenta Math. Extra Volume ICM **I**, 467–486 (1998)

[72] P. Shor, Introduction to quantum algorithms. AMS Proc. Symp. Appl. Math. **58**, 17 (2002)

[73] P. Shor, Why haven't more quantum algorithms been found? J. ACM **50**(1), 87–90 (2003)

[74] D.R. Simon, On the power of quantum computation. SIAM J. Comput. **26**(5), 1471–1483 (1997)

[75] V. Strassen, Einige Resultate über Berechnungskomplexität. Jahresber. Dtsch. Math. Ver. **78**, 1–84 (1976/1997)

[76] W. Trappe, L. Washington, *Introduction to Cryptography with Coding Theory*, 2nd edn. (Prentice-Hall, Englewood Cliffs, 2006)

[77] L.M.K. Vandersypen, M. Steffen, G. Breyta, C.S. Tannoni, M.H. Sherwood, I.L. Chuang, Experimental realization of Shor's quantum factoring algorithm using nuclear magnetic resonance. Nature **414**, 883–887 (2001)

[78] R. Van Meter, K.M. Itoh, Fast quantum modular exponentiation. Phys. Rev. A **71**, 052320 (2005)

[79] R. Van Meter, W.J. Munro, K. Nemoto, Architecture of a quantum milticomputer implementing Shor's algorithm, in *Theory of Quantum Computation, Communication and Cryptography*, ed. by Y. Kawano, M. Mosca. Lecture Note in Computer Science, vol. 5106 (Springer, Berlin, 2008), pp. 105–114

[80] U.V. Vazirani, On the power of quantum computation. Phil. Trans. R. Soc. Lond. **A356**, 1759–1768 (1998)

[81] U.V. Vazirani, Fourier transforms and quantum computation, in *Proceedings of Theoretical Aspects of Computer Science* (Springer, Berlin, 2000), pp. 208–220

[82] U.V. Vazirani, A survey of quantum complexity theory. AMS Proc. Symp. Appl. Math. **58**, 28 (2002)

[83] J. Watrous, in *Quantum Computational Complexity*. Encyclopedia of Complexity and System Science (Springer, New York, 2009), pp. 7174–7201

[84] H. Wiener, Cryptanalysis of short RSA secret exponents. IEEE Trans. Inf. Theor. **36**(3), 553–558 (1990)

[85] C.P. Williams, *Explorations in Quantum Computation*, 2nd edn. (Springer, New York, 2011)

[86] S.Y. Yan, *Cryptanalyic Attacks on RSA* (Springer, Berlin, 2008)

[87] S.Y. Yan, in *Primality Testing and Integer Factorization in Public-Key Cryptography*. Advances in Information Security, vol. 11, 2nd edn. (Springer, New York, 2009)

[88] N.S. Yanofsky, M.A. Mannucci, *Quantum Computing for Computer Scientists* (Cambridge University Press, Cambridge, 2008)

[89] A.C. Yao, Quantum circuit complexity, in *Proceedings of Foundations of Computer Science* (IEEE, New York, 1993), pp. 352–361

[90] C. Zalka, Fast versions of Shor's quantum factoring algorithm. LANA e-print quant-ph 9806084, p. 37 (1998)

3. Quantum Attacks on DLP-Based Cryptosystems

Try a hard problem. You may not solve it, but you will prove something else.

JOHN LITTLEWOOD (1885–1977)
Great British Mathematician

In this chapter, we shall first formally define the discrete logarithm problem (DLP) and some classical solutions to DLP. Then we shall discuss the DLP-based cryptographic systems and protocols whose security depends on the infeasibility of the DLP problem. Finally, we shall discuss a quantum approach to attacking both the DLP problem and the DLP-based cryptography.

3.1 DLP and Classic Solutions to DLP

Basic Concepts of DLP

Definition 3.1. The *DLP* can be described as follows:

$$\left.\begin{array}{ll} \text{Input}: & a,b,n \in \mathbb{Z}^+ \\ \text{Output}: & x \in \mathbb{N} \text{ with } a^x \equiv b \ (\text{mod } n) \\ & \text{if such an } x \text{ exists,} \end{array}\right\} \qquad (3.1)$$

where the modulus n can either be a composite or a prime.

According to Adleman in [1], the Russian mathematician Bouniakowsky developed a clever algorithm to solve the congruence $a^x \equiv b \ (\text{mod } n)$, with the asymptotic complexity $\mathcal{O}(n)$ in 1870. Despite its long history, no efficient

S.Y. Yan, *Quantum Attacks on Public-Key Cryptosystems*,
DOI 10.1007/978-1-4419-7722-9_3,
© Springer Science+Business Media, LLC 2013

algorithm has ever emerged for the DLP. It is believed to be extremely hard, and harder than the integer factorization problem (IFP) even in the average case. The best known algorithm for DLP at present, using NFS and due to Gordon [21], requires an expected running time

$$\mathcal{O}\left(\exp\left(c(\log n)^{1/3}(\log\log n)^{2/3}\right)\right).$$

There are essentially three different categories of algorithms in use for computing discrete logarithms:

1. Algorithms that work for arbitrary groups, that is, those that do not exploit any specific properties of groups; Shanks' baby-step giant-step method, Pollard's ρ-method (an analog of Pollard's ρ-factoring method), and the λ-method (also known as wild and tame Kangaroos) [42] are in this category.

2. Algorithms that work well in finite groups for which the order of the groups has no large prime factors; more specifically, algorithms that work for groups with smooth orders. A positive integer is called *smooth* if it has no large prime factors; it is called *y-smooth* if it has no large prime factors exceeding y. The well-known Silver–Pohlig–Hellman algorithm based on the Chinese Remainder Theorem is in this category.

3. Algorithms that exploit methods for representing group elements as products of elements from a relatively small set (also making use of the Chinese Remainder Theorem); the typical algorithms in this category are Adleman's index calculus algorithm and Gordon's NFS algorithm.

In the sections that follow, we shall introduce the basic ideas of each of these three categories; more specifically, we shall introduce Shanks' baby-step giant-step algorithm, Silver–Pohlig–Hellman algorithm, Adleman's index calculus Algorithm, as well as Gordon's NFS algorithm for computing discrete logarithms.

Shanks' Baby-Step Giant-Step Algorithm

Let G be a finite cyclic group of order n, a a generator of G and $b \in G$. The *obvious* algorithm for computing successive powers of a until b is found takes $\mathcal{O}(n)$ group operations. For example, to compute $x = \log_2 15 \pmod{19}$, we compute $2^x \bmod 19$ for $x = 0, 1, 2, \ldots, 19 - 1$ until $2^x \bmod 19 = 15$ for some x is found, that is:

x	0	1	2	3	4	5	6	7	8	9	10	11
a^x	1	2	4	8	16	13	7	14	9	18	17	15

So $\log_2 15 \pmod{19} = 11$. It is clear that when n is large, the algorithm is inefficient. In this section, we introduce a type of square root algorithm, called the baby-step giant-step algorithm, for taking discrete logarithms, which is better than the above-mentioned *obvious* algorithm. The algorithm, due to Daniel Shanks (1917–1996), works on arbitrary groups [52].

Let $m = \lfloor \sqrt{n} \rfloor$. The baby-step giant-step algorithm is based on the observation that if $x = \log_a b$, then we can uniquely write $x = i + jm$, where $0 \leqslant i, j < m$. For example, if $11 = \log_2 15 \bmod 19$, then $a = 2$, $b = 15$, $m = 5$, so we can write $11 = i + 5j$ for $0 \leqslant i, j < m$. Clearly here $i = 1$ and $j = 2$ so we have $11 = 1 + 5 \cdot 2$. Similarly, for $14 = \log_2 6 \bmod 19$, we can write $14 = 4 + 5 \cdot 2$, for $17 = \log_2 10 \bmod 19$, we can write $17 = 2 + 5 \cdot 3$, etc. The following is a description of the algorithm:

Algorithm 3.1 (Shanks' baby-step giant-step algorithm). This algorithm computes the discrete logarithm x of y to the base a, modulo n, such that $y = a^x \pmod{n}$:

[1] (Initialization) Computes $s = \lfloor \sqrt{n} \rfloor$.

[2] (Computing the baby-step) Compute the first sequence (list), denoted by S, of pairs (ya^r, r), $r = 0, 1, 2, 3, \ldots, s - 1$:

$$S = \{(y, 0), (ya, 1), (ya^2, 2), (ya^3, 3), \ldots, (ya^{s-1}, s - 1) \bmod n\} \qquad (3.2)$$

and sort S by ya^r, the first element of the pairs in S.

[3] (Computing the giant-step) Compute the second sequence (list), denoted by T, of pairs (a^{ts}, ts), $t = 1, 2, 3, \ldots, s$:

$$T = \{(a^s, 1), (a^{2s}, 2), (a^{3s}, 3), \ldots, (a^{s^2}, s) \bmod n\} \qquad (3.3)$$

and sort T by a^{ts}, the first element of the pairs in T.

[4] (Searching, comparing, and computing) Search both lists S and T for a match $ya^r = a^{ts}$ with ya^r in S and a^{ts} in T and then compute $x = ts - r$. This x is the required value of $\log_a y \pmod{n}$.

This algorithm requires a table with $\mathcal{O}(m)$ entries ($m = \lfloor \sqrt{n} \rfloor$, where n is the modulus). Using a sorting algorithm, we can sort both the lists S and T in $\mathcal{O}(m \log m)$ operations. Thus, this gives an algorithm for computing discrete logarithms that uses $\mathcal{O}(\sqrt{n} \log n)$ time and space for $\mathcal{O}(\sqrt{n})$ group elements. Note that Shanks' idea was originally for computing the order of a group element g in the group G, but here we use his idea to compute discrete logarithms. Note also that although this algorithm works on arbitrary groups, if the order of a group is larger than 10^{40}, it will be infeasible.

Example 3.1. Suppose we wish to compute the discrete logarithm $x = \log_2 6 \bmod 19$ such that $6 = 2^x \bmod 19$. According to Algorithm 3.1, we perform the following computations:

[1] $y = 6$, $a = 2$ and $n = 19$, $s = \lfloor \sqrt{19} \rfloor = 4$.

[2] Computing the baby-step:

$$
\begin{aligned}
S &= \{(y, 0), (ya, 1), (ya^2, 2), (ya^3, 3) \mod 19\} \\
&= \{(6, 0), (6 \cdot 2, 1), (6 \cdot 2^2, 2), (6 \cdot 2^3, 3) \mod 19\} \\
&= \{(6, 0), (12, 1), (5, 2), (10, 3)\} \\
&= \{(5, 2), (6, 0), (10, 3), (12, 1)\}.
\end{aligned}
$$

[3] Computing the giant-step:

$$
\begin{aligned}
T &= \{(a^s, s), (a^{2s}, 2s), (a^{3s}, 3s), (a^{4s}, 4s) \mod 19\} \\
&= \{(2^4, 4), (2^8, 8), (2^{12}, 12), (2^{16}, 16) \mod 19\} \\
&= \{(16, 4), (9, 8), (11, 12), (5, 16)\} \\
&= \{(5, 16), (9, 8), (11, 12), (16, 4)\}.
\end{aligned}
$$

[4] Matching and computing: The number 5 is the common value of the first element in pairs of both lists S and T with $r = 2$ and $st = 16$, so $x = st - r = 16 - 2 = 14$. That is, $\log_2 6 \pmod{19} = 14$ or, equivalently, $2^{14} \pmod{19} = 6$.

Example 3.2. Suppose now we wish to find the discrete logarithm $x = \log_{59} 67 \mod 113$, such that $67 = 59^x \mod 113$. Again by Algorithm 3.1, we have:

[1] $y = 67$, $a = 59$ and $n = 113$, $s = \lfloor \sqrt{113} \rfloor = 10$.

[2] Computing the baby-step:

$$
\begin{aligned}
S &= \{(y, 0), (ya, 1), (ya^2, 2), (ya^3, 3), \ldots, (ya^9, 9) \mod 113\} \\
&= \{(67, 0), (67 \cdot 59, 1), (67 \cdot 59^2, 2), (67 \cdot 59^3, 3), (67 \cdot 59^4, 4), \\
&\quad\ (67 \cdot 59^5, 5), (67 \cdot 59^6, 6), (67 \cdot 59^7, 7), (67 \cdot 59^8, 8), \\
&\quad\ (67 \cdot 59^9, 9) \mod 113\} \\
&= \{(67, 0), (111, 1), (108, 2), (44, 3), (110, 4), (49, 5), (66, 6), \\
&\quad\ (52, 7), (17, 8), (99, 9)\} \\
&= \{(17, 8), (44, 3), (49, 5), (52, 7), (66, 6), (67, 0), (99, 9), \\
&\quad\ (108, 2), (110, 4), (111, 1)\}.
\end{aligned}
$$

[3] Computing the giant-step:

$$
\begin{aligned}
T &= \{(a^s, s), (a^{2s}, ss), (a^{3s}, 3s), \ldots (a^{10s}, 10s) \bmod 113\} \\
&= \{(59^{10}, 10), (59^{2 \cdot 10}, 2 \cdot 10), (59^{3 \cdot 10}, 3 \cdot 10), (59^{4 \cdot 10}, 4 \cdot 10), \\
&\quad\ (59^{5 \cdot 10}, 5 \cdot 10), (59^{6 \cdot 10}, 6 \cdot 10), (59^{7 \cdot 10}, 7 \cdot 10), (59^{8 \cdot 10}, 8 \cdot 10), \\
&\quad\ (59^{9 \cdot 10}, 9 \cdot 10) \bmod 113\} \\
&= \{(72, 10), (99, 20), (9, 30), (83, 40), (100, 50), (81, 60), \\
&\quad\ (69, 70), (109, 80), (51, 90), (56, 100)\} \\
&= \{(9, 30), (51, 90), (56, 100), (69, 70), (72, 10), (81, 60), (83, 40), \\
&\quad\ (99, 20), (100, 50), (109, 80)\}.
\end{aligned}
$$

[4] Matching and computing: The number 99 is the common value of the first element in pairs of both lists S and T with $r = 9$ and $st = 20$, so $x = st - r = 20 - 9 = 11$. That is, $\log_{59} 67 \pmod{113} = 11$ or, equivalently, $59^{11} \pmod{113} = 67$.

Shanks' baby-step giant-step algorithm is a type of *square root method* for computing discrete logarithms. In 1978 Pollard also gave two other types of square root methods, namely, the ρ-method and the λ-method for taking discrete logarithms. Pollard's methods are probabilistic but remove the necessity of precomputing the lists S and T, as with Shanks' baby-step giant-step method. Again, Pollard's algorithm requires $\mathcal{O}(n)$ group operations and hence is infeasible if the order of the group G is larger than 10^{40}.

Silver–Pohlig–Hellman Algorithm

In 1978, Pohlig and Hellman proposed an important special algorithm, now widely known as the Silver–Pohlig–Hellman algorithm for computing discrete logarithms over $\mathrm{GF}(q)$ with $\mathcal{O}(\sqrt{p})$ operations and a comparable amount of storage, where p is the largest prime factor of $q - 1$. Pohlig and Hellman showed that if

$$q - 1 = \prod_{i=1}^{k} p_i^{\alpha_i}, \tag{3.4}$$

where the p_i are distinct primes and the α_i are natural numbers, and if $r_1, \ldots, r_k$ are any real numbers with $0 \leqslant r_i \leqslant 1$, then logarithms over $\mathrm{GF}(q)$ can be computed in

$$\mathcal{O}\left(\sum_{i=1}^{k} \left(\log q + p_i^{1-r_i} \left(1 + \log p_i^{r_i}\right)\right)\right) \tag{3.5}$$

field operations, using

$$\mathcal{O}\left(\log q \sum_{i=1}^{k} \left(1 + p_i^{r_i}\right)\right) \tag{3.6}$$

bits of memory, provided that a precomputation requiring

$$\mathcal{O}\left(\sum_{i=1}^{k} p_i^{r_i} \log p_i^{r_i} + \log q\right) \tag{3.7}$$

field operations is performed first. This algorithm is very efficient if q is "smooth", i.e., all the prime factors of $q - 1$ are small. We shall give a brief description of the algorithm as follows:

Algorithm 3.2 (Silver–Pohlig–Hellman Algorithm). This   algorithm computes the discrete logarithm $x = \log_a b \bmod q$:

[1] Factor $q - 1$ into its prime factorization form:

$$q - 1 = \prod_{i=1}^{k} p_1^{\alpha_1} p_2^{\alpha_2} \cdots p_k^{\alpha_k}.$$

[2] Precompute the table $r_{p_i, j}$ for a given field:

$$r_{p_i, j} = a^{j(q-1)/p_i} \bmod q, \quad 0 \leqslant j < p_i. \tag{3.8}$$

This only needs to be done once for any given field.

[3] Compute the discrete logarithm of b to the base a modulo q, i.e., compute $x = \log_a b \bmod q$:

[3-1] Use an idea similar to that in the baby-step giant-step algorithm to find the individual discrete logarithms $x \bmod p_i^{\alpha_i}$: To compute $x \bmod p_i^{\alpha_i}$, we consider the representation of this number to the base p_i:

$$x \bmod p_i^{\alpha_i} = x_0 + x_1 p_i + \cdots + x_{\alpha_i - 1} p_i^{\alpha_i - 1}, \tag{3.9}$$

where $0 \leqslant x_n < p_i - 1$.

(a) To find x_0, we compute $b^{(q-1)/p_i}$ which equals $r_{p_i, j}$ for some j, and set $x_0 = j$ for which

$$b^{(q-1)/p_i} \bmod q = r_{p_i, j}.$$

This is possible because

$$b^{(q-1)/p_i} \equiv a^{x(q-1)/p} \equiv a^{x_0(q-1)/p} \bmod q = r_{p_i, x_0}.$$

(b) To find x_1, compute $b_1 = ba^{-x_0}$. If

$$b_1^{(q-1)/p_i^2} \bmod q = r_{p_i, j},$$

then set $x_1 = j$. This is possible because

$$b_1^{(q-1)/p_i^2} \equiv a^{(x-x_0)(q-1)/p_i^2} \equiv a^{(x_1+x_2 p_i + \cdots)(q-1)/p_i}$$
$$\equiv a^{x_1(q-1)/p} \bmod q = r_{p_i,x_1}.$$

(c) To obtain x_2, consider the number $b_2 = b a^{-x_0 - x_1 p_i}$ and compute

$$b_2^{(q-1)/p_i^3} \bmod q.$$

The procedure is carried on inductively to find all $x_0, x_1, \ldots, x_{\alpha_i - 1}$.

[3-2] Use the Chinese Remainder Theorem to find the unique value of x from the congruences $x \bmod p_i^{\alpha_i}$.

We now give an example of how the above algorithm works:

Example 3.3. Suppose we wish to compute the discrete logarithm $x = \log_2 62 \bmod 181$. Now we have $a = 2$, $b = 62$, and $q = 181$ (2 is a generator of $\mathbb{F}_{181}^*$). We follow the computation steps described in the above algorithm:

[1] Factor $q - 1$ into its prime factorization form:

$$180 = 2^2 \cdot 3^2 \cdot 5.$$

[2] Use the following formula to precompute the table $r_{p_i,j}$ for the given field $\mathbb{F}_{181}^*$:

$$r_{p_i,j} = a^{j(q-1)/p_i} \bmod q, \quad 0 \leqslant j < p_i.$$

This only needs to be done once for this field.

(a) Compute

$$r_{p_1,j} = a^{j(q-1)/p_1} \bmod q = 2^{90j} \bmod 181 \text{ for } 0 \leqslant j < p_1 = 2:$$

$$r_{2,0} = 2^{90 \cdot 0} \bmod 181 = 1,$$
$$r_{2,1} = 2^{90 \cdot 1} \bmod 181 = 180.$$

(b) Compute

$$r_{p_2,j} = a^{j(q-1)/p_2} \bmod q = 2^{60j} \bmod 181 \text{ for } 0 \leqslant j < p_2 = 3:$$

$$r_{3,0} = 2^{60 \cdot 0} \bmod 181 = 1,$$
$$r_{3,1} = 2^{60 \cdot 1} \bmod 181 = 48,$$
$$r_{3,2} = 2^{60 \cdot 2} \bmod 181 = 132.$$

(c) Compute

$$r_{p_3,j} = a^{j(q-1)/p_3} \bmod q = 2^{36j} \bmod 181 \text{ for } 0 \leqslant j < p_3 = 5:$$

$$r_{5,0} = 2^{36\cdot 0} \bmod 181 = 1,$$

$$r_{5,1} = 2^{36\cdot 1} \bmod 181 = 59,$$

$$r_{5,2} = 2^{36\cdot 2} \bmod 181 = 42,$$

$$r_{5,3} = 2^{36\cdot 3} \bmod 181 = 125,$$

$$r_{5,4} = 2^{36\cdot 4} \bmod 181 = 135.$$

Construct the $r_{p_i,j}$ table as follows:

p_i	j				
	0	1	2	3	4
2	1	180			
3	1	48	132		
5	1	59	42	125	135

This table is manageable if all p_i are small.

[3] Compute the discrete logarithm of 62 to the base 2 modulo 181, that is, compute $x = \log_2 62 \bmod 181$. Here $a = 2$ and $b = 62$:

[3-1] Find the individual discrete logarithms $x \bmod p_i^{\alpha_i}$ using

$$x \bmod p_i^{\alpha_i} = x_0 + x_1 p_i + \cdots + x_{\alpha_i - 1} p_i^{\alpha_i - 1}, \quad 0 \leqslant x_n < p_i - 1.$$

(a-1) Find the discrete logarithms $x \bmod p_1^{\alpha_1}$, i.e., $x \bmod 2^2$:

$$x \bmod 181 \Longleftrightarrow x \bmod 2^2 = x_0 + 2x_1.$$

(i) To find x_0, we compute

$$b^{(q-1)/p_1} \bmod q = 62^{180/2} \bmod 181 = 1 = r_{p_1,j} = r_{2,0};$$

hence $x_0 = 0$.

(ii) To find x_1, compute first $b_1 = ba^{-x_0} = b = 62$ and then compute

$$b_1^{(q-1)/p_1^2} \bmod q = 62^{180/4} \bmod 181 = 1 = r_{p_1,j} = r_{2,0};$$

hence $x_1 = 0$. So

$$x \bmod 2^2 = x_0 + 2x_1 \Longrightarrow x \bmod 4 = 0.$$

(a-2) Find the discrete logarithms $x \bmod p_2^{\alpha_2}$, that is, $x \bmod 3^2$:

$$x \bmod 181 \Longleftrightarrow x \bmod 3^2 = x_0 + 2x_1.$$

(i) To find x_0, we compute

$$b^{(q-1)/p_2} \bmod q = 62^{180/3} \bmod 181 = 48 = r_{p_2,j} = r_{3,1};$$

hence $x_0 = 1$.

(ii) To find x_1, compute first $b_1 = ba^{-x_0} = 62 \cdot 2^{-1} = 31$ and then compute

$$b_1^{(q-1)/p_2^2} \bmod q = 31^{180/3^2} \bmod 181 = 1 = r_{p_2,j} = r_{3,0};$$

hence $x_1 = 0$. So

$$x \bmod 3^2 = x_0 + 2x_1 \Longrightarrow x \bmod 9 = 1.$$

(a-3) Find the discrete logarithms $x \bmod p_3^{\alpha_3}$, that is, $x \bmod 5^1$:

$$x \bmod 181 \Longleftrightarrow x \bmod 5^1 = x_0.$$

To find x_0, we compute

$$b^{(q-1)/p_3} \bmod q = 62^{180/5} \bmod 181 = 1 = r_{p_3,j} = r_{5,0};$$

hence $x_0 = 0$. So we conclude that

$$x \bmod 5 = x_0 \Longrightarrow x \bmod 5 = 0.$$

[3-2] Find the x in

$$x \bmod 181,$$

such that

$$\begin{cases} x \bmod 4 = 0, \\ x \bmod 9 = 1, \\ x \bmod 5 = 0. \end{cases}$$

To do this, we just use the Chinese Remainder Theorem to solve the following system of congruences:

$$\begin{cases} x \equiv 0 \pmod{4}, \\ x \equiv 1 \pmod{9}, \\ x \equiv 0 \pmod{5}. \end{cases}$$

The unique value of x for this system of congruences is $x = 100$. (This can be easily done by using, for example, the Maple function chrem([0, 1, 0], [4,9, 5]).) So the value of x in the congruence $x \bmod 181$ is 100. Hence, $x = \log_2 62 = 100$.

ρ-Method for DLP

We have seen that the Pollard ρ-method [41] can be used to solve the IFP problem. We shall see that there is a corresponding algorithm of ρ for solving the DLP problem [42], which has the same expected running time as the baby-step and giant-step, but which requires a negligible amount of storage. Assume we wish to find x such that

$$\alpha^x \equiv \beta \pmod{n}.$$

Note that we assume the order of the element α in the multiplicative group $\mathbb{Z}_n^*$ is r. In ρ for DLP, the group $G = \mathbb{Z}_n^*$ is partitioned into three sets G_1, G_2, and G_3 of roughly equal size. Define a sequence of group elements $\{x_i\}$: $x_0, x_1, x_2, x_3, \cdots$ as follows:

$$
\begin{cases}
x_0 = 1, \\
x_{i+1} = f(x_i) = \begin{cases}
\beta \cdot x_i, & \text{if } x_i \in G_1, \\
x_i^2, & \text{if } x_i \in G_1, \\
\alpha \cdot x_i, & \text{if } x_i \in G_1,
\end{cases}
\end{cases}
\tag{3.10}
$$

for $i \geqslant 0$. This sequence in turn defines two sequences of integers $\{a_i\}$ and $\{b_i\}$ as follows:

$$
\begin{cases}
a_0 = 0, \\
a_{i+1} = \begin{cases}
a_i, & \text{if } x_i \in G_1, \\
2a_i, & \text{if } x_i \in G_1, \\
a_i + 1, & \text{if } x_i \in G_1,
\end{cases}
\end{cases}
\tag{3.11}
$$

and

$$
\begin{cases}
b_0 = 0, \\
b_{i+1} = \begin{cases}
b_i + 1, & \text{if } x_i \in G_1, \\
2b_i, & \text{if } x_i \in G_2, \\
b_i, & \text{if } x_i \in G_3,
\end{cases}
\end{cases}
\tag{3.12}
$$

Just the same as ρ for IFP, we find two group elements x_i and x_{2i} such that $x_i = x_{2i}$. Hence,

$$\alpha^{a_i} \beta^{b_i} = \alpha^{2a_i} \beta^{2b_i}.$$

Therefore,

$$\beta^{b_i - 2b_i} = \alpha^{2a_i - a_i}. \tag{3.13}$$

By taking logarithm to the base α of both sides in (3.13), we get

$$x = \log_\alpha \beta \equiv \frac{2a_i - a_i}{b_i - 2b_i} \pmod{r}, \tag{3.14}$$

provided that $b_i \not\equiv 2b_i \pmod{n}$. The corresponding ρ algorithm may be described as follows.

Algorithm 3.3 (ρ for DLP). This algorithm tries to find x such that

$$\alpha^x \equiv \beta \ (\mathrm{mod}\ n).$$

Set $x_0 = 1, a_0 = 0, b_0 = 0$
For $i = 1, 2, 3, \cdots$ do
    Using (3.10), (3.11) and (3.12) to compute (x_i, a_i, b_i) and
    (x_{2i}, a_{2i}, b_{2i})
    If $x_i = x_{2i}$, do
        Set $r \leftarrow b_i - b_{2i} \bmod n$
        If $r = 0$ terminate the algorithm with failure
            else compute $x \equiv r^{-1}(a_{2i} - a_i) \ (\mathrm{mod}\ n)$
        output x

Example 3.4. Solve x such that

$$89^x \equiv 618 \ (\mathrm{mod}\ 809).$$

Let G_1, G_2, G_3 be as follows:

$$G_1 = \{x \in \mathbb{Z}_{809} : \ x \equiv 1 \ (\mathrm{mod}\ 3)\},$$
$$G_2 = \{x \in \mathbb{Z}_{809} : \ x \equiv 0 \ (\mathrm{mod}\ 3)\},$$
$$G_3 = \{x \in \mathbb{Z}_{809} : \ x \equiv 2 \ (\mathrm{mod}\ 3)\}.$$

For $i = 1, 2, 3, \cdots$, we calculate (x_i, a_i, b_i) and (x_{2i}, a_{2i}, b_{2i}) until $x_i = x_{2i}$ as
follows:

i	$(\mathbf{x_i}, a_i, b_i)$	$(\mathbf{x_{2i}}, a_{2i}, b_{2i})$
1	$(681, 0, 1)$	$(76, 0, 2)$
2	$(76, 0, 2)$	$(113, 0, 4)$
3	$(46, 0, 3)$	$(488, 1, 5)$
4	$(113, 0, 4)$	$(605, 4, 10)$
5	$(349, 1, 4)$	$(422, 5, 11)$
6	$(488, 1, 5)$	$(683, 7, 11)$
7	$(555, 2, 5)$	$(451, 8, 12)$
8	$(605, 4, 10)$	$(344, 9, 13)$
9	$(451, 5, 10)$	$(112, 11, 13)$
10	$(\mathbf{422}, 5, 11)$	$(\mathbf{422}, 11, 15)$

At $i = 10$, I have found a match that

$$x_{10} = x_{20} = 422.$$

Since the order of 89 in $\mathbb{Z}_{809}^*$ is 101, we have

$$x \equiv \frac{a_{2i} - a_i}{b_i - b_{2i}},$$

$$\equiv \frac{11 - 5}{11 - 15}$$

$$\equiv 49 \ (\mathrm{mod}\ 101).$$

Clearly,
$$89^{49} \equiv 618 \ (\text{mod} \ 809).$$

Index Calculus Algorithm

In 1979, Adleman [1] proposed a general purpose, subexponential algorithm for computing discrete logarithms, called the *index calculus method*, with the following expected running time:

$$\mathcal{O}\left(\exp\left(c\sqrt{\log n \log \log n}\right)\right).$$

The index calculus is, in fact, a wide range of methods, including CFRAC, QS, and NFS for IFP. In what follows, we discuss a variant of Adleman's index calculus for DLP in $(\mathbb{Z}/p\mathbb{Z})^*$.

Algorithm 3.4 (Index calculus for DLP). This algorithm tries to find an integer k such that

$$k \equiv \log_\beta \alpha \ (\text{mod} \ p) \ \ \text{or} \ \ \alpha \equiv \beta^k \ (\text{mod} \ p).$$

[1] Precomputation

[1-1] (Choose Factor Base) Select a factor base Γ, consisting of the first m prime numbers,
$$\Gamma = \{p_1, p_2, \ldots, p_m\},$$
with $p_m \leqslant B$, the bound of the factor base.

[1-2] (Compute $\beta^e \bmod p$) Randomly choose a set of exponent $e \leqslant p-2$, compute $\beta^e \bmod p$, and factor it as a product of prime powers.

[1-3] (Smoothness) Collect only those relations $\beta^e \bmod p$ that are smooth with respect to B. That is,

$$\beta^e \bmod p = \prod_{i=1}^{m} p_i{}^{e_i}, e_i \geqslant 0. \tag{3.15}$$

When such relations exist, get

$$e \equiv \sum_{j=1}^{m} e_j \log_\beta p_j \ (\text{mod} \ p-1). \tag{3.16}$$

[1-4] (Repeat) Repeat [1-3] to find at least m such e in order to find m relations as in (3.16) and solve $\log_\beta p_j$ for $j = 1, 2, \ldots, m$.

[2] Compute $k \equiv \log_\beta \alpha \pmod{p}$

[2-1] For each e in (3.16), determine the value of $\log_\beta p_j$ for $j = 1, 2, \ldots, m$ by solving the m modular linear equations with unknown $\log_\beta p_j$.

[2-2] (Compute $\alpha\beta^r \bmod p$) Randomly choose exponent $r \leqslant p - 2$ and compute $\alpha\beta^r \bmod p$.

[2-3] (Factor $\alpha\beta^r \bmod p$ over Γ)

$$\alpha\beta^r \bmod p = \prod_{j=1}^{m} p_j^{r_i}, r_j \geqslant 0. \tag{3.17}$$

If (3.17) is unsuccessful, go back to Step [2-2]. If it is successful, then

$$\log_\beta \alpha \equiv -r + \sum_{j=1}^{m} r_j \log_\beta p_j. \tag{3.18}$$

Example 3.5 (Index calculus for DLP). Find

$$x \equiv \log_{22} 4 \pmod{3361}$$

such that

$$4 \equiv 22^x \pmod{3361}.$$

[1] Precomputation

[1-1] (Choose Factor Base) Select a factor base Γ, consisting of the first 4 prime numbers,

$$\Gamma = \{2, 3, 5, 7\},$$

with $p_4 \leqslant 7$, the bound of the factor base.

[1-2] (Compute $22^e \bmod 3361$) Randomly choose a set of exponent $e \leqslant 3359$, compute $22^e \bmod 3361$, and factor it as a product of prime powers:

$$22^{48} \equiv 2^5 \cdot 3^2 \pmod{3361},$$
$$22^{100} \equiv 2^6 \cdot 7 \pmod{3361},$$
$$22^{186} \equiv 2^9 \cdot 5 \pmod{3361},$$
$$22^{2986} \equiv 2^3 \cdot 3 \cdot 5^2 \pmod{3361}.$$

[1-3] (Smoothness) The above four relations are smooth with respect to $B = 7$. Thus,

$$48 \equiv 5 \log_{22} 2 + 2 \log_{22} 3 \pmod{3360},$$
$$100 \equiv 6 \log_{22} 2 + \log_{22} 7 \pmod{3360},$$
$$186 \equiv 9 \log_{22} 2 + \log_{22} 5 \pmod{3360},$$
$$2986 \equiv 3 \log_{22} 2 + \log_{22} 3 + 2 \log_{22} 5 \pmod{3360}.$$

[2] Compute $k \equiv \log_\beta \alpha \pmod{p}$

    [2-1] Compute
$$\log_{22} 2 \equiv 1100 \pmod{3360},$$
$$\log_{22} 3 \equiv 2314 \pmod{3360},$$
$$\log_{22} 5 \equiv 366 \pmod{3360},$$
$$\log_{22} 7 \equiv 220 \pmod{3360}.$$

    [2-2] (Compute $4 \cdot 22^r \bmod p$) Randomly choose exponent $r = 754 \leqslant$ 3659 and compute $4 \cdot 22^{754} \bmod 3361$.

    [2-3] (Factor $4 \cdot 22^{754} \bmod 3361$ over Γ)

$$4 \cdot 22^{754} \equiv 2 \cdot 3^2 \cdot 5 \cdot 7 \pmod{3361}.$$

Thus,

$$
\begin{aligned}
\log_{22} 4 &\equiv -754 + \log_{22} 2 + 2\log_{22} 3 + \log_{22} 5 + \log_{22} 7 \\
&\equiv 2200.
\end{aligned}
$$

That is,
$$22^{2200} \equiv 4 \pmod{3361}.$$

Example 3.6. Find $k \equiv \log_{11} 7 \pmod{29}$ such that $\beta^k \equiv 11 \pmod{29}$.

[1] (Factor Base) Let the factor base $\Gamma = \{2, 3, 5\}$.

[2] (Compute and Factor $\beta^e \bmod p$) Randomly choose $e < p$, compute, and factor $\beta^e \bmod p = 11^e \bmod 29$ as follows:

    (1) $11^2 \equiv 5 \pmod{29}$ (success),

    (2) $11^3 \equiv 2 \cdot 13 \pmod{29}$ (fail),

    (3) $11^5 \equiv 2 \cdot 7 \pmod{29}$ (fail)

    (4) $11^6 \equiv 3^2 \pmod{29}$ (success),

    (5) $11^7 \equiv 2^3 \cdot 3 \pmod{29}$ (success),

    (6) $11^9 \equiv 2 \cdot 7 \pmod{29}$ (success).

[3] (Solve the systems of congruences for the quantities $\log_\beta p_i$)

    (1) $\log_{11} 5 \equiv 2 \pmod{28}$,

    (4) $\log_{11} 3 \equiv 3 \pmod{28}$,

    (6) $\log_{11} 2 \equiv 9 \pmod{28}$,

    (5) $2 \cdot \log_{11} 2 + \log_{11} 3 \equiv 7 \pmod{28}$,

          $\log_{11} 3 \equiv 17 \pmod{28}$.

[4] (Compute and Factor $\alpha\beta^e \bmod p$) Randomly choose $e < p$, compute, and factor $\alpha\beta^e \bmod p = 7 \cdot 11^e \bmod 29$ as follows:

$$7 \cdot 11 \equiv 19 \ (\text{mod } 29) \qquad (\text{fail}),$$
$$7 \cdot 11^2 \equiv 2 \cdot 3 \ (\text{mod } 29) \qquad (\text{success}).$$

Thus,

$$\log_{11} 7 \equiv \log_{11} 2 + \log_{11} 3 - 2 \equiv 24 \ (\text{mod } 28).$$

This is true since

$$11^{24} \equiv 7 \ (\text{mod } 29).$$

For more than 10 years since its invention, Adleman's method and its variants were the fastest algorithms for computing discrete logarithms. But the situation changed when Gordon [21] in 1993 proposed an algorithm for computing discrete logarithms in $\mathrm{GF}(p)$. Gordon's algorithm is based on the number field sieve (NFS) for integer factorization, with the heuristic expected running time

$$\mathcal{O}\left(\exp\left(c(\log p)^{1/3} (\log\log p)^{2/3} \right) \right),$$

the same as that used in factoring. The algorithm can be briefly described as follows:

Algorithm 3.5 (Gordon's NFS). This algorithm computes the discrete logarithm x such that $a^x \equiv b \ (\text{mod } p)$ with input a, b, p, where a and b are generators and p is prime:

[1] (Precomputation): Find the discrete logarithms of a factor base of small rational primes, which must only be done once for a given p.

[2] (Compute individual logarithms): Find the logarithm for each $b \in \mathbb{F}_p$ by finding the logarithms of a number of "medium-sized" primes.

[3] (Compute the final logarithm): Combine all the individual logarithms (by using the Chinese Remainder Theorem) to find the logarithm of b.

Interested readers are referred to Gordon's paper [21] for more detailed information. Note also that Gordon, with coauthor McCurley [22], discussed some implementation issues of massively parallel computations of discrete logarithms over $\mathrm{GF}(2^n)$.

Exercises and Problems for Sect. 3.1

1. Use the exhaustive method to find the following discrete logarithms k over $\mathbb{Z}_{1009}^*$, if exist:

(a) $k \equiv \log_3 57 \pmod{1009}$.

(b) $k \equiv \log_{11} 57 \pmod{1009}$.

(c) $k \equiv \log_3 20 \pmod{1009}$.

2. Use the baby-step giant-step algorithm to compute the following discrete logarithms k:
   (a) $k \equiv \log_5 96 \pmod{317}$.

   (b) $k \equiv \log_{37} 15 \pmod{123}$.

   (c) $k \equiv \log_5 57105961 \pmod{58231351}$.

3. Use Silver–Pohlig–Hellman algorithm to solve the discrete logarithms k:
   (a) $3^k \equiv 2 \pmod{65537}$.

   (b) $5^k \equiv 57105961 \pmod{58231351}$.

   (c) $k \equiv \log_5 57105961 \pmod{58231351}$.

4. Use Pollard's ρ method to find the discrete logarithms k such that
   (a)
   $$2^k \equiv 228 \pmod{383}.$$

   (b) $5^k \equiv 3 \pmod{2017}$.

5. Let the factor base $\Gamma = \{2, 3, 5, 7\}$. Use the index calculus method to find the discrete logarithm k:
   $$k \equiv \log_2 37 \pmod{131}.$$

6. Use the index calculus with factor base $\Gamma = (2, 3, 5, 7, 11)$ to solve the DLP problem
   $$k \equiv \log_7 13 \pmod{2039}.$$

7. Let

$$
\begin{aligned}
p &= 31415926535897932384626433832795028841971693993751058209 \\
  &= 74944592307816406286208998628034825342117067982148086513 \\
  &= 28230664709384460955058223172535940812848 1237299,
\end{aligned}
$$

$$x = 2,$$

$$
\begin{aligned}
y &= 27182818284590452353602874713526624977572470936999595749 \\
  &\quad 66976277240766303535475945713821785251664274274663919 32 \\
  &\quad 0030599218174135966290435729003342952605956 30738.
\end{aligned}
$$

(a) Use Gordon's index calculus method (Algorithm 3.5) to compute the k such that
$$y \equiv x^k \pmod{p}.$$

(b) Verify that if your k is as follows:
82989716465034897051864680264075784402496146932312647219853184518689598402644834266625285046612688143761738165394262430753767931963671156105352608242351366559643075376793196367115610535260824235136655964307537679319636711561053526082423513665596

3.2 DLP-Based Cryptography

Basic Idea of DLP-Based Cryptography

Just the same as IFP, DLP is also a computationally infeasible problem in number theory, so we can use DLP to construct cryptographic system that should be secure and unbreakable in polynomial time.

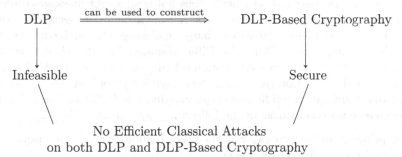

In this section, we introduce various popular cryptographic systems whose security relies on the infeasibility of the DLP problem, including the first public-key system—the DHM key-exchange system and the most influential ElGamal encryption and digital signature systems.

The Diffie–Hellman–Merkle Key-Exchange Protocol

Diffie and Hellman [16] in 1976 proposed for the first time the concept and idea of public-key cryptography and the first public-key system based on the infeasible DLP. Their system is not a public-key cryptographic system but a public-key distribution system based on Merkle's seminal work in 1978 [35] (Fig. 3.1 shows the photo of Merkle, Hellman and Diffie in the 1970s).

Figure 3.1. Merkle, Hellman and Diffie (Courtesy of Prof. Hellman)

Such a public-key distribution scheme does not send secret messages directly but rather allows the two parties to agree on a common secret key over public networks to be used later in exchanging messages through conventional secret-key cryptography. Thus, the Diffie–Hellman–Merkle scheme has the nice property that a very fast encryption scheme such as DES or AES can be used for actual encryption (just using the agreed key), yet it still enjoys one of the main advantages of public-key cryptography. The Diffie–Hellman–Merkle key-exchange protocol works in the following way (see also Fig. 3.2):

1. A prime q and a generator g are made public (assume all users have agreed upon a finite group over a fixed finite field $\mathbb{F}_q$).

2. Alice chooses a random number $a \in \{1, 2, \ldots, q-1\}$ and sends $g^a \bmod q$ to Bob.

3. Bob chooses a random number $b \in \{1, 2, \ldots, q-1\}$ and sends $g^b \bmod q$ to Alice.

4. Alice and Bob both compute $g^{ab} \bmod q$ and use this as a private key for future communications.

Clearly, an eavesdropper has g, q, $g^a \bmod q$ and $g^b \bmod q$, so if he can take discrete logarithms, he can calculate $g^{ab} \bmod q$ and understand the communications. That is, if the eavesdropper can use his knowledge of g, q, $g^a \bmod q$ and $g^b \bmod q$ to recover the integer a, then he can easily break the Diffie–Hellman–Merkle system. So the security of the Diffie–Hellman–Merkle system is based on the following assumption:

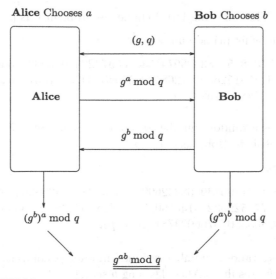

Figure 3.2. DHM key-exchange protocol

Diffie–Hellman–Merkle assumption: It is computationally infeasible to compute $g^{ab} \bmod q$ from $g, q, g^a \bmod q$ and $g^b \bmod q$. That is,

$$\{g, q, g^a \bmod q, g^b \bmod q\} \xrightarrow{\text{hard to find}} \{g^{ab} \bmod q\}.$$

The Diffie–Hellman–Merkle assumption, in turn, depends on the following DLP assumption, i.e.,

$$\{g, q, g^a \bmod q\} \xrightarrow{\text{hard to find}} \{a\},$$

or

$$\{g, q, g^b \bmod q\} \xrightarrow{\text{hard to find}} \{b\}.$$

In theory, there could be a way to use the knowledge of $g^a \bmod q$ and $g^b \bmod q$ to find $g^{ab} \bmod q$. But at present, we simply cannot imagine a way to go from $g^a \bmod q$ and $g^b \bmod q$ to $g^{ab} \bmod q$ without essentially solving the following DLP:

$$\{g, q, g^a \bmod q\} \xrightarrow{\text{find}} \{a\},$$

or

$$\{g, q, g^b \bmod q\} \xrightarrow{\text{find}} \{b\}.$$

If either a or b can be find efficiently, then DHM can be broken easily, since

$$\{g, q, b, g^a \bmod q\} \xrightarrow{\text{easy to find}} \{(g^a)^b \equiv g^{ab} \ (\bmod \ q)\}$$

or

$$\{g, q, a, g^b \bmod q\} \xrightarrow{\text{easy to find}} \{(g^b)^a \equiv g^{ab} \ (\bmod \ q)\}.$$

Example 3.7. The following DHM challenge problem was proposed in [33].

[1] Let p be following prime number:

$p = 20470627038553283805974453516697427480360839434012345\underline{9}\_$
$69579867459152659137268522951065284733970579762207550\underline{5}\_$
$069831043486651682279.$

[2] Alice chooses a random number a modulo p, computes 7^a (mod p), and sends the result to Bob, keeping a secret.

[3] Bob receives

$7^a \equiv 127402180119973946824269244334322849749382042586931\underline{62}\_$
$165455773529032291467909599868186097881304659516645\underline{54}\_$
58144280588076766033781 (mod p).

[4] Bob chooses a random number residue b modulo p, computes 7^b (mod p), and sends the result to Alice, keeping b secret.

[5] Alice receives

$7^b \equiv 180162285287453102444782834836799895015967046695346\underline{69}\_$
$731302512173405995377205847595817691062538069210165\underline{18}\_$
48662362137934026803049 (mod p).

[6] Now both Alice and Bob can compute the private-key 7^{ab} (mod p).

McCurley offered a prize of \$100 in 1989 to the first person or group to find the private key constructed from the above communication.

Example 3.8. McCurley's 129-digit discrete logarithm challenge was actually solved on 25 January 1998 using the NFS method, by the two German computer scientists, Weber at the Institut für Techno-und Wirtschaftsmathematik in Kaiserslautern and Denny at the Debis IT Security Services in Bonn [68]. Their solution to McCurley's DLP problem is as follows:

$a \equiv 38127280411190014138078391507929634193998643551018670\underline{285}\_$
$56137516504552396692940392210217251405327092887266394\underline{263}\_$
70063532797740808 (mod p),
$(7^b)^a \equiv 61858690859651883273593331665203790426798764306952\underline{1}\_$
$713459146222184952599815614487782075749218290977\underline{740}\_$
$83387918504579467\underline{49734}.$

As we have already mentioned earlier the Diffie–Hellman–Merkle scheme is not intended to be used for actual secure communications but for key exchanges. There are, however, several other cryptosystems based on discrete logarithms that can be used for secure message transmissions.

ElGamal Cryptography

In 1985, ElGamal [19], a Ph.D. student of Hellman at Stanford then, proposed the first DLP-based public-key cryptosystem, since the plaintext M can be recovered by taking the following discrete logarithms:

$$M \equiv \log_{M^e} M \ (\text{mod } n).$$

The ElGamal cryptosystem can be described as follows (see also Fig. 3.3).

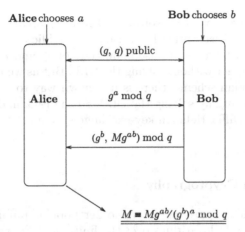

Figure 3.3. ElGamal cryptography

1. A prime q and a generator $g \in \mathbb{F}_q^*$ are made public.

2. Alice chooses at random a private integer

$$a \in \{1, 2, \ldots, q-1\}. \tag{3.19}$$

   This a is the private decryption key. The public encryption key is $\{g, q, g^a \bmod q\}$.

3. Suppose now Bob wishes to send a message to Alice. He chooses a random number $b \in \{1, 2, \ldots, q-1\}$ and sends Alice the following pair of elements of $\mathbb{F}_q$:

$$(g^b, \ Mg^{ab})$$

   where M is the message.

4. Since Alice knows the private decryption key a, she can recover M from this pair by computing $g^{ab} \pmod{q}$ and dividing this result into the second element. That is,

$$M \equiv Mg^{ab}/(g^b)^a \ (\text{mod } q).$$

5. Cryptanalysis: Find the private a by solving the DLP problem

$$a \equiv \log_g x \ (\text{mod } q - 1)$$

such that

$$x \equiv g^a \ (\text{mod } q).$$

Remark 3.1. Anyone who can solve the DLP in $\mathbb{F}_q$ breaks the cryptosystem by finding the secret decryption key a from the public encryption key g^a. In theory, there could be a way to use knowledge of g^a and g^b to find g^{ab} and hence break the cipher without solving the DLP. But as we have already seen in the Diffie–Hellman scheme, there is no known way to go from g^a and g^b to g^{ab} without essentially solving the DLP. So the ElGamal cryptosystem is equivalent to the Diffie–Hellman key-exchange system.

Massey–Omura Cryptography

The Massey–Omura cryptosystem is another popular public-key cryptosystem based on discrete logarithms over the finite field $\mathbb{F}_q$, with $p = p^r$ prime power. It was proposed by James Massey and Jim K. Omura in 1982 [32] as a possible improvement over Shamir's original three-pass cryptographic protocol developed around 1980, in which the sender and the receiver do not exchange any keys; however, the protocol does require the sender and receiver to have two private keys for encrypting and decrypting messages. Thus, the Massey–Omura cryptosystem works in the following steps (see Fig. 3.4):

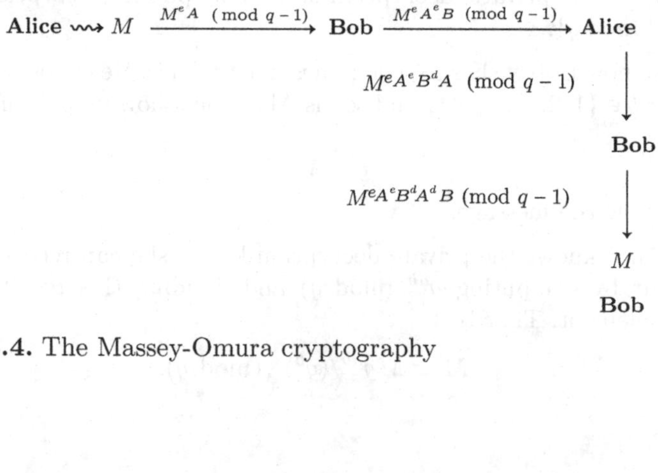

Alice $\rightsquigarrow M$ $\xrightarrow{\ M^e A \ (\text{mod } q-1)\ }$ Bob $\xrightarrow{\ M^e A^e B \ (\text{mod } q-1)\ }$ Alice

$$M^e A^e B^d A \ (\text{mod } q-1) \qquad \downarrow$$

Bob

$$M^e A^e B^d A^d B \ (\text{mod } q-1) \qquad \downarrow$$

$$M$$

Bob

Figure 3.4. The Massey-Omura cryptography

1. All the users have agreed upon a finite group over a fixed finite field $\mathbb{F}_q$ with q a prime power.

2. Each user secretly selects a random integer e between 0 and $q-1$ such that $\gcd(e, q-1) = 1$ and computes $d = e^{-1} \bmod (q-1)$ by using the extended Euclidean algorithm. At the end of this step, Alice gets (e_A, d_A) and Bib gets (e_B, d_B).

3. Now suppose that user Alice wishes to send a secure message M to user Bob, then they follow the following procedure:

   (a) Alice first sends M^{e_A} to Bob.

   (b) On receiving Alice's message, Bob sends $M^{e_A e_B}$ back to Alice (note that at this point, Bob cannot read Alice's message M).

   (c) Alice sends $M^{e_A e_B d_A} = M^{e_B}$ to Bob.

   (d) Bob then computes $M^{d_B e_B} = M$ and hence recovers Alice's original message M.

4. Cryptanalysis: Eve shall be hard to find M from the three-pass protocol between Alice and Bob unless she can solve the DLP involved efficiently.

The Massey–Omura cryptosystem may also be described in detail as follows.

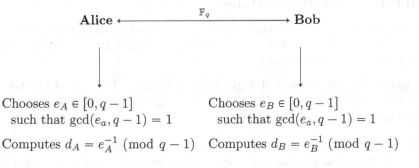

$$M \equiv M^{e_A e_B d_A d_B} \pmod{q-1}$$

Example 3.9. Let

$$p = 80000000000000001239,$$
$$M = 20210519040125 \text{ (Tuesday)},$$
$$e_A = 6654873997,$$
$$e_B = 7658494001.$$

Then

$$d_A \equiv \tfrac{1}{e_A} \equiv 70094446778448900393 \ (\text{mod } p-1),$$
$$d_B \equiv \tfrac{1}{e_B} \equiv 14252518250422012923 \ (\text{mod } p-1),$$
$$M^{e_A} \equiv 56964332403383118724 \ (\text{mod } p),$$
$$M^{e_A e_B} \equiv 37671804887541585024 \bmod p,$$
$$M^{e_A e_B d_A} \equiv 50551151743565447865 \bmod p,$$
$$M^{e_A e_B d_A d_B} \equiv 20210519040125 \ (\text{mod } p),$$
$$\downarrow$$
$$M$$

DLP-Based Digital Signatures

The ElGamal's cryptosystem [19] can also be used for digital signatures; the security of such a signature scheme depends on the intractability of discrete logarithms over a finite field.

Algorithm 3.6 (ElGamal Signature Scheme). This algorithm tries to generate digital signature $S = (a, b)$ for message m. Suppose that Alice wishes to send a signed message to Bob.

[1] [ElGamal key generation] Alice does the following:

    [1-1] Choose a prime p and two random integers g and x, such that both g and x are less than p.

    [1-2] Compute $y \equiv g^x \ (\text{mod } p)$.

    [1-3] Make (y, g, p) public (both g and p can be shared among a group of users), but keep x as a secret.

[2] [ElGamal signature generation] Alice does the following:

    [2-1] Choose at random an integers k such that $\gcd(k, p-1) = 1$.

    [2-2] Compute

$$\left. \begin{array}{l} a \equiv g^k \ (\text{mod } p), \\[2mm] b \equiv k^{-1}(m - xa) \ (\text{mod } (p-1)). \end{array} \right\} \qquad (3.20)$$

Now Alice has generated the signature (a, b). She must keep the random integer, k, as secret.

[3] [ElGamal signature verification] To verify Alice's signature, Bob confirms that

$$y^a a^b \equiv g^m \ (\text{mod } p). \tag{3.21}$$

In August 1991, the US government's National Institute of Standards and Technology (NIST) proposed an algorithm for digital signatures. The algorithm is known as DSA, for digital signature algorithm. The DSA has become the US Federal Information Processing Standard 186 (FIPS 186). It is called the digital signature standard (DSS) [64] and is the first digital signature scheme recognized by any government. The role of DSA/DSS is expected to be analogous to that of the Data Encryption Standard (DES). The DSA/DSS is similar to a signature scheme proposed by Schnorr; it is also similar to a signature scheme of ElGamal. The DSA is intended for use in electronic mail, electronic funds transfer, electronic data interchange, software distribution, data storage, and other applications which require data integrity assurance and data authentication. The DSA/DSS consists of two main processes:

[1] Signature generation (using the private key)

[2] Signature verification (using the public-key)

A one-way hash function is used in the signature generation process to obtain a condensed version of data, called a message digest. The message digest is then signed. The digital signature is sent to the intended receiver along with the signed data (often called the message). The receiver of the message and the signature verifies the signature by using the sender's public-key. The same hash function must also be used in the verification process. In what follows, we shall give the formal specifications of the DSA/DSS.

Algorithm 3.7 (DSA). This is a variation of ElGamal signature scheme. It generates a signature $S = (r, s)$ for the message m.

[1] [DSA key generation] To generate the DSA key, the sender performs the following:

   [1-1] Find a 512-bit prime p (which will be public).

   [1-2] Find a 160-bit prime q dividing evenly into $p - 1$ (which will be public).

   [1-3] Generate an element $g \in \mathbb{Z}/p\mathbb{Z}$ whose multiplicative order is q, i.e., $g^q \equiv 1 \ (\text{mod } p)$.

   [1-4] Find a one-way function H mapping messages into 160-bit values.

   [1-5] Choose a secret key x, with $0 < x < q$,

   [1-6] Choose a public key y, where $y \equiv g^x \ (\text{mod } p)$.

Clearly, the secret x is the discrete logarithm of y, modulo p, to the base g.

[2] [DSA Signature Generation] To sign the message m, the sender produces his signature as (r, s), by selecting a random integer $k \in \mathbb{Z}/q\mathbb{Z}$ and computing

$$\left. \begin{array}{l} r \equiv \left(g^k \ (\mathrm{mod} \ p)\right) \ (\mathrm{mod} \ q), \\[2mm] s \equiv k^{-1}(H(m) + xr) \ (\mathrm{mod} \ q). \end{array} \right\} \qquad (3.22)$$

[3] [DSA Signature Verification] To verify the signature (r, s) for the message m from the sender, the receiver first computes

$$t \equiv s^{-1} \ (\mathrm{mod} \ q) \qquad (3.23)$$

and then accepts the signature as valid if the following congruence holds:

$$r \equiv \left(g^{H(m)t} y^{rt} \ (\mathrm{mod} \ p)\right) \ (\mathrm{mod} \ q). \qquad (3.24)$$

If the congruence (3.24) does not hold, then the message either may have been incorrectly signed or may have been signed by an impostor. In this case, the message is considered to be invalid.

There are, however, many responses solicited by the (US) Association of Computing Machinery (ACM), positive and negative, to the NIST's DSA. Some positive aspects of the DSA include:

1. The US government has finally recognized the utility and the usefulness of public-key cryptography. In fact, the DSA is the only signature algorithm that has been publicly proposed by any government.

2. The DSA is based on reasonable familiar number-theoretic concepts, and it is especially useful to the financial services industry.

3. Signatures in DSA are relatively short (only 320 bits), and the key generation process can be performed very efficiently.

4. When signing, the computation of r can be done even before the message m is available, in a "precomputation" step.

While some negative aspects of the DSA include:

1. The DSA does not include key exchanges and cannot be used for key distribution and encryption.

2. The key size in DSA is too short; it is restricted to a 512-bit modulus or key size, which is too short and should be increased to at least 1024 bits.

3. The DSA is not compatible with existing international standards; for example, the international standards organizations such as ISO, CCITT, and SWIFT all have accepted the RSA as a standard.

Nevertheless, the DSA is the only one publicly known government DSS.

Exercises and Problems for Sect. 3.2

1. In McCurley's DLP problem, we have

$$7^b \equiv 18016228528745310244478283483679989501596704669534669\_$$
$$731302512173405995377205847595817691062538069210165518\_$$
$$48662362137934026803049 \pmod{p},$$
$$p = 20470627038553283380597445351669742748036083943401234 59\_$$
$$695798674591526591372685229510652847339705797622075505\_$$
$$069831043486651682279.$$

(a) Find the discrete logarithm b.

(b) Compute $(7^a)^b \bmod p$.

(c) Verify if your result $(7^a)^b \bmod p$ agrees to Weber and Denny's result, i.e., check if $(7^a)^b \equiv (7^b)^a \pmod{p}$.

2. Let the DHM parameters be as follows:

$$p \quad = \quad 10000000000000000000000000000000000000000000000000000\_$$
$$00000000000000000002047062703855328380597445351669742\_$$
$$74803608394340123459695798674591526591372685229510652\_$$
$$8473397057976220755050698310434866516828 89,$$

$$13^x \quad \equiv \quad 10851945926748930321536897787511601536291411551215963\_$$
$$735797413754705002845778243766666788726776122805935 69,\_$$
$$523266148125732037472098621361064920285476333105415 81\_$$
$$3024411985737741571370874416352991514462 6 \pmod{p},$$

$$13^y \quad \equiv \quad 52200208400156523080484387248076760362198322255017014\_$$
$$267256873745866707749922777188091986977849828727835 84\_$$
$$838294594895654776487332569999727232277536865712330 58\_$$
$$30747697800417855036551198719274264122371 \pmod{p}.$$

(a) Find the discrete logarithm x.

(b) Find the discrete logarithm y.

(c) Compute $(13^x)^y \pmod{p}$.

(d) Compute $(13^y)^x \pmod{p}$.

3. In ElGamal cryptosystem, Alice makes (p, g, g^a) public with p prime p:

p = 100000000000000000000000000000000000000000000000000000000_
000000000000000000002047062703855328380597445351669742_
748046083943401234596957986745915265913726852295106520_
84733970579762207550506983104348665168328 1

g = 137

g^a ≡ 15219266397668101959283316151426320683674451858111063_
45767690506157955692567935509944285656491006943855496_
14388735928661950422196794512676225936419253780225375_
37252639984353500071774531090027331523676

where $a \in \{1, 2, \cdots, p\}$ must be kept as a secret. Now Bob can send an encrypted message $C = (g^b, Mg^{ab})$ to Alice by using her public-key information, where

g^b ≡ 59547675601458322302365604133720220696052746940473 3_
55046049744137914374142183634043230653659070816467 4_
62466636904384382001528769925211730081006654249356 4_
12826389882146691842217779072611842406374051259

Mg^{ab} ≡ 49587861882815113830430418447664907530237264453603 2_
94479849527736721533557707864314686330644624599660 5_
60087834147651129038106201491085560126484952668340 8_
83323263742065525535496981642865216817002959760

(a) Find the discrete logarithm a, and compute $(g^b)^a \bmod p$.

(b) Find the discrete logarithm b, and compute $(g^a)^b \bmod p$.

(c) Decode the ciphertext C by computer either

$$M \equiv Mg^{ab}/(g^b)^a \pmod{p}$$

or

$$M \equiv Mg^{ab}/(g^a)^b \pmod{p}.$$

4. Let

$p = 14197,$
$(e_A, d_A) = (13, 13105),$
$(e_B, d_B) = (17, 6681),$
$M = 1511 \text{ (OK)}.$

Find
$$M^{e_A} \bmod p,$$
$$M^{e_A e_B} \bmod p,$$
$$M^{e_A e_B d_A} \bmod p,$$
$$M^{e_A e_B d_A d_B} \bmod p$$

and check if $M \equiv M^{e_A e_B d_A d_B} \pmod{p}$.

5. Let
$$p = 20000000000000002559,$$
$$M = 2015140426251511811 \text{ (To New York)},$$
$$e_A = 6654873997,$$
$$e_B = 7658494001.$$

(a) Find
$$d_A \equiv 1/e_A \pmod{p-1},$$
$$d_B \equiv 1/e_B \pmod{p-1}.$$

(b) Find
$$M^{e_A} \bmod p,$$
$$M^{e_A e_B} \bmod p,$$
$$M^{e_A e_B d_A} \bmod p,$$
$$M^{e_A e_B d_A d_B} \bmod p.$$

(c) Check if $M \equiv M^{e_A e_B d_A d_B} \pmod{p}$.

6. Suppose, in ElGamal cryptosystem, the random number k is chosen to sign two different messages. Let
$$\begin{aligned} b_1 &\equiv k^{-1}(m_1 - xa) \pmod{(p-1)}, \\ b_2 &\equiv k^{-1}(m_2 - xa) \pmod{(p-1)} \end{aligned}$$

where
$$a \equiv g^k \pmod{p}.$$

(a) Show that k can be computed from
$$(b_1 - b_2)k \equiv (m_1 - m_2) \pmod{(p-1)}.$$

(b) Show that the private key x can be determined from the knowledge of k.

7. Show that breaking DHM key-exchange scheme or any DLP-based cryptosystem is generally equivalent to solving the DLP problem.

3.3 Quantum Attack on DLP and DLP-Based Cryptography

Relationships Between DLP and DLP-Based Cryptography

As can be seen, DLP is a conjectured (i.e., unproved) infeasible problem in computational number theory; this would imply that the cryptographic system-based DLP is secure and unbreakable in polynomial time:

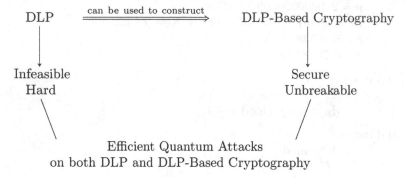

Thus, anyone who can solve DLP can break DLP-based cryptography. With this regard, solving DLP is equivalent to breaking DLP-based cryptography. As everybody knows at present, no efficient algorithm is known for solving DLP, therefore, no efficient algorithm for cracking DLP-based cryptography. However, Shor [53] showed that DLP can be solved in $\mathcal{BQP}$, where $\mathcal{BQP}$ is the class of problem that is efficiently solvable in polynomial time on a quantum Turing machine, just in the same idea of quantum factoring attacks on IFP-based cryptography:

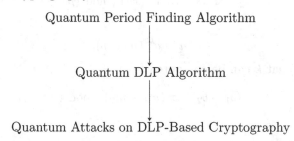

Hence, all DLP-based cryptographic systems can be broken in polynomial time on a quantum computer.

Level of Difficulty for Different DLP

There are three main types of DLP problems, according to the level of the difficulty to solve them:

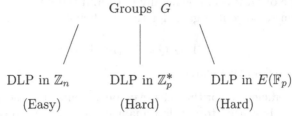

$$\text{Groups } G$$

DLP in $\mathbb{Z}_n$	DLP in $\mathbb{Z}_p^*$	DLP in $E(\mathbb{F}_p)$
(Easy)	(Hard)	(Hard)

1. DLP in additive group $G = \mathbb{Z}_n$ is *easy* to compute: Let us consider the additive (cyclic) group $G = \mathbb{Z}_{100}$ of order 100. Find

$$n \equiv \log_3 17 \pmod{100}$$

   such that

$$3n \equiv 17 \pmod{100}.$$

   This type of DLP can be computed in polynomial time by using Euclid's algorithm for multiplicative inverse as follows:

$$\begin{aligned} n &= \frac{1}{3} \cdot 17 \\ &= 67 \cdot 17 \\ &= 39. \end{aligned}$$

2. DLP in multiplicative group $G = \mathbb{Z}_p^*$ is *hard* to compute: Let us consider the multiplicative (cyclic) group $G = \mathbb{Z}_{101}^*$ of order 100. Find

$$n \equiv \log_3 17 \pmod{101}$$

   such that

$$3^n \equiv 17 \pmod{101}.$$

   This type of DLP is generally hard and there is no polynomial-time algorithm to solve it. Of course, for this artificially small example, one can find

$$\log_3 17 \equiv 70 \pmod{101}$$

   easily by exhaustive search.

3. DLP in elliptic curve group $G = E(\mathbb{F}_p)$ is also *hard* to compute. Consider the elliptic curve over a finite field as follows:

$$E \backslash \mathbb{F}_{101} : y^2 \equiv x^3 + 7x + 12 \pmod{101}$$

where $\{P(-1,2), Q(31,86)\} \in E(\mathbb{F}_{101})$. Find $k \equiv \log_P Q \pmod{101}$ such that

$$\log_P Q \equiv k \pmod{101}.$$

This type of DLP is also generally hard and there is no polynomial-time algorithm to solve it. Again, for this artificially small example, one can find

$$\log_P Q \equiv 78 \pmod{101}$$

easily by exhaustive search.

Since no efficient algorithm has ever been found for the last two types of DLP, there is a need to develop quantum algorithms to solve the problems. This is exactly the motivation for quantum algorithms and quantum computers.

Idea of Quantum DLP Attacks

Recall that in DLP, we wish to find k in

$$g^r \equiv x \pmod{p}, \qquad (3.25)$$

where g is a generator in the multiplicative group $\mathbb{Z}_p^*$. We assume the order of g in $\mathbb{Z}_p^*$ is known to be k, that is,

$$g^k \equiv 1 \pmod{p}. \qquad (3.26)$$

Notice first that in quantum factoring algorithm, we try to find

$$g^r \equiv 1 \pmod{p},$$

where r is the order of g in $\mathbb{P}_{p-1}$. In quantum discrete logarithm algorithm, we try to find

$$g^r \equiv x \pmod{p}, \qquad (3.27)$$

where r is discrete logarithm to the base g in $\mathbb{P}_{p-1}$, that is,

$$r \equiv \log_g x \pmod{p-1}. \qquad (3.28)$$

The definitions of r in the two quantum algorithms are different. However, since

$$g^r \equiv x \pmod{p},$$

we can define a 2-variable function (just the same as $f(a) = g^r \equiv 1 \pmod{p}$ in quantum algorithm):

$$f(a,b) = g^a x^{-b} \equiv 1 \pmod{p} \qquad (3.29)$$

such that

$$a - br \equiv k \pmod{p}, \tag{3.30}$$

which can be so, because

$$
\begin{aligned}
g^a x^{-b} &\equiv g^a (g^r)^{-b} \tag{3.31} \\
&\equiv g^a g^{-br} \tag{3.32} \\
&\equiv g^{a-br} \tag{3.33} \\
&\equiv g^k \pmod{p}. \tag{3.34}
\end{aligned}
$$

Thus, in quantum discrete logarithm algorithm, we essentially need to solve r in

$$r \equiv (a - k)b^{-1} \pmod{p - 1}, \tag{3.35}$$

which is, in turn, just an inverse problem. Shor [53] shows that the quantum algorithm can solve r in polynomial time. Of course, if p is smooth (i.e., $p - 1$ must have small prime factors), then DLP in $\mathbb{Z}_p^*$ can already be solved in polynomial time by Pohlig–Hellman algorithm [40] (we call this case as an easy case of DLP). However, for general p, there is still no classical polynomial time for DLP (we call this case as a hard case of DLP). In what follows, we shall first discuss the easy case and then the hard case of the quantum DLP attacks.

Easy Case of Quantum DLP Attack

The easy case of the quantum DLP attack is basically the quantum analog or quantum version of the Pohlig–Hellman method for DLP. Recall that to find the discrete logarithm r in

$$g^r \equiv x \pmod{p}$$

where g is a generator of the multiplicative group $\mathbb{Z}^*$ and p is a prime with $p - 1$ smooth, Pohlig–Hellman method can solve the problem efficiently in polynomial time on a classical computer. It looks no advantage to use quantum computers to solve this particular easy, smooth case of DLP. However, it is a good exercise to show that a quantum computer can solve a problem just the same as a classical computer.

Algorithm 3.8 (Quantum algorithm for general DLP). Given $g, x \in \mathbb{N}$ and p prime. This algorithm will find the integer r such that $g^r \equiv x \pmod{p}$ if r exists. It uses three quantum registers.

[1] Beginning with the initial state

$$|\Psi_0\rangle = |0\rangle|0\rangle, \tag{3.36}$$

choose numbers a and b modulo $p-1$ uniformly and perform a Fourier transform modulo $p-1$, denoted by A_{p-1}. So the state of the machine after this step is

$$
\begin{aligned}
|\Psi_1\rangle &= \frac{1}{\sqrt{p-1}} \sum_{a=0}^{p-2} |a\rangle \cdot \frac{1}{\sqrt{p-1}} \sum_{b=0}^{p-2} |b\rangle \\
&= \frac{1}{p-1} \sum_{a=0}^{p-2} \sum_{b=0}^{p-2} |a,\, b\rangle.
\end{aligned}
\tag{3.37}
$$

[2] Compute $g^a x^{-b} \pmod{p}$ reversibly (the values of a and b must be kept on the tape (just memory, in terms of quantum Turing machine, we call tape). This leaves the quantum computer in the state $|\Psi_2\rangle$:

$$|\Psi_2\rangle = \frac{1}{p-1} \sum_{a=0}^{p-2} \sum_{b=0}^{p-2} |a,\, b,\, g^a x^{-b} \pmod{p}\rangle. \tag{3.38}$$

[3] Use the Fourier transform A_{p-1} to map $|a\rangle \rightarrow |c\rangle$ with probability amplitude

$$\sqrt{\frac{1}{p-1}} \exp\left(\frac{2\pi i a c}{p-1}\right)$$

and $|b\rangle \rightarrow |d\rangle$ with probability amplitude

$$\sqrt{\frac{1}{p-1}} \exp\left(\frac{2\pi i b d}{p-1}\right).$$

Thus, the state $|a, b\rangle$ will be changed to the state:

$$\frac{1}{(p-1)^2} \sum_{a,c=0}^{p-2} \sum_{b,d=0}^{p-2} \exp\left(\frac{2\pi i}{p-1}(ac+bd)\right) |c,\, d\rangle. \tag{3.39}$$

This leaves the machine in the state $|\Psi_2\rangle$:

$$|\Psi_3\rangle = \frac{1}{(p-1)^2} \sum_{a,b,c,d=0}^{p-2} \exp\left(\frac{2\pi i}{p-1}(ac+bd)\right) |c,\, d,\, g^a x^{-b} \pmod{p}\rangle. \tag{3.40}$$

[4] Observe the state of the quantum computer and extract the required information. The probability of observing a state $\left| c,\ d,\ g^k \pmod{p} \right\rangle$ is

$$\text{Prob}(c, d, g^k) = \left| \frac{1}{(p-1)^2} \sum_{\substack{a,b \\ a-rb\equiv k \pmod{p-1}}} \exp\left(\frac{2\pi i}{p-1}(ac+bd) \right) \right|^2$$

(3.41)

where the sum is over all (a, b) such that

$$a - rb \equiv k \pmod{p-1}. \tag{3.42}$$

[5] Substituting

$$a \equiv k + rb \pmod{p-1} \tag{3.43}$$

in (3.42), we get

$$\text{Prob}(c, d, g^k) = \left| \frac{1}{(p-1)^2} \sum_{b}^{p-2} \exp\left(\frac{2\pi i}{p-1}(kc + b(d+rc)) \right) \right|^2 \tag{3.44}$$

Notice that if $d + rc \not\equiv 0 \pmod{p-1}$, then the probability is 0. Thus, the probability $\neq 0$ if and only if $d + rc \equiv 0 \pmod{p-1}$, that is,

$$r \equiv -dc^{-1} \pmod{p-1}. \tag{3.45}$$

[6] As our computation has produced a random c and the corresponding $d \equiv -rc \pmod{p-1}$. Thus, if $\gcd(c, p-1) = 1$, then we can find r by finding the multiplicative inverse of c using Euclid's algorithm. More importantly, the chance that $\gcd(c, p-1) = 1$ is

$$\frac{\phi(p-1)}{p-1} > \frac{1}{\log p},$$

in fact,

$$\liminf \frac{\phi(p-1)}{p-1} \approx \frac{e^{-\gamma}}{\log\log p}.$$

So we only need a number of experiments that is polynomial in $\log p$ to obtain r with high probability.

General Case of Quantum DLP Attack

We have just showed that quantum computers can solve a computational problem, namely, the special case of DLP, just the same as classical computer. However, a quantum computer may also be able to solve a computational problem efficiently in polynomial time, namely, the general case of DLP, that cannot be solved efficiently in polynomial time on a classical computer. Here is the quantum algorithm.

Recall that the special case DLP is based on the fact that $p-1$ is smooth. In the general case, we remove this restriction by choosing a random smooth q such that $p \leqslant q \leqslant 2p$; it can be shown that such a q can be found in polynomial time such that no prime power larger than $c \log q$ divides q for some constant c independent of p.

Algorithm 3.9 (Quantum algorithm for general DLP). Let g be a generator of $\mathbb{Z}_p^*$, $x \in \mathbb{Z}_p$. This algorithm will find the integer r such that $g^r \equiv x \pmod{p}$.

[1] Choose a random smooth number q such that $p \leqslant q \leqslant 2p$. Note that we do not require $p-1$ to be smooth.

[2] Just the same as the special case, choose numbers a and b modulo $p-1$ uniformly and perform a Fourier transform modulo $p-1$. This leaves the quantum computer in the state $|\Psi_1\rangle$:

$$|\Psi_1\rangle = \frac{1}{p-1} \sum_{a=0}^{p-2} \sum_{b=0}^{p-2} |a,\ b \pmod{p}\rangle. \tag{3.46}$$

[2] Compute $g^a x^{-b} \bmod p$ reversibly. This leaves the quantum computer in the state $|\Psi_2\rangle$:

$$|\Psi_2\rangle = \frac{1}{p-1} \sum_{a=0}^{p-2} \sum_{b=0}^{p-2} |a,\ b,\ g^a x^{-b} \pmod{p}\rangle. \tag{3.47}$$

[3] Use the Fourier transform A_q to map $|a\rangle \rightarrow |c\rangle$ with the probability amplitude

$$\frac{1}{\sqrt{q}} \exp\left(\frac{2\pi iac}{q}\right)$$

and $|b\rangle \rightarrow |d\rangle$ with probability amplitude

$$\frac{1}{\sqrt{q}} \exp\left(\frac{2\pi ibd}{q}\right).$$

Thus, the state $|a,b\rangle$ will be changed to the state:

$$\frac{1}{p-1} \sum_{c=0}^{p-2} \sum_{d=0}^{p-2} \exp\left(\frac{2\pi i}{q}(ac+bd)\right) |c,\ d\rangle. \tag{3.48}$$

This leaves the machine in the state $|\Psi_3\rangle$:

$$|\Psi_3\rangle = \frac{1}{(p-1)q} \sum_{a,b=0}^{p-2} \sum_{c,d=0}^{q-1} \exp\left(\frac{2\pi i}{q}(ac+bd)\right) |c, \ d, \ g^a x^{-b} \ (\mathrm{mod} \ p)\rangle.$$

$$(3.49)$$

[4] Observe the state of the quantum computer and extract the required information. The probability of observing a state $|c, \ d, \ g^k \ (\mathrm{mod} \ p)\rangle$ is almost the same as the special case:

$$\mathrm{Prob}(c, d, g^k) = \left| \frac{1}{(p-1)q} \sum_{\substack{a,b \\ a-rb \equiv k \ (\mathrm{mod} \ p-1)}} \exp\left(\frac{2\pi i}{q}(ac+bd)\right) \right|^2 \qquad (3.50)$$

where the sum is over all (a, b) such that

$$a - rb \equiv k \ (\mathrm{mod} \ p-1). \qquad (3.51)$$

[5] Use the relation

$$a \equiv k + br - (p-1)\left\lfloor \frac{br+k}{p-1} \right\rfloor \qquad (3.52)$$

and substitute in (3.50) to obtain the amplitude:

$$\frac{1}{(p-1)q} \sum_{b=0}^{p-2} \exp\left(\frac{2\pi i}{q}\left(brc + kc + bd - c(p-1)\left\lfloor \frac{br+k}{p-1} \right\rfloor\right)\right), \qquad (3.53)$$

so that the sum of (3.50) becomes

$$\left| \frac{1}{(p-1)q} \sum_{b=0}^{p-2} \exp\left(\frac{2\pi i}{q}\left(brc + kc + bd - c(p-1)\left\lfloor \frac{br+k}{p-1} \right\rfloor\right)\right) \right|^2,$$

$$(3.54)$$

which is the probability of observing the state $|c, \ d, \ g^k \ (\mathrm{mod} \ p)\rangle$.

[6] It can be shown that a certain pair of values of c, d occurs with high probability and satisfies the bound

$$\left| rc + d - \frac{r}{p-1}(c(p-1) \ \mathrm{mod} \ q) \right| \leqslant \frac{1}{2}. \qquad (3.55)$$

Once such a pair c, d can be found, r can be deduced, as r is the only unknown in

$$\left| d + \frac{r(c(p-1) - c(p-1) \ \mathrm{mod} \ q)}{p-1} \right| \leqslant \frac{1}{2}. \qquad (3.56)$$

Notice also that

$$q \mid (c(p-1) - c(p-1) \ \mathrm{mod} \ q).$$

Then dividing both sides by q, we get

$$\left| \frac{d}{q} - \frac{rl}{p-1} \right| \leq \frac{1}{2q}. \tag{3.57}$$

To find r, just round $\frac{d}{q}$ to the closest multiple of $p-1$, denoted by $\frac{m}{p-1}$, and then compute r from

$$\frac{m}{p-1} = \frac{rl}{p-1}. \tag{3.58}$$

That is,

$$r = \frac{m}{l}. \tag{3.59}$$

Exercises and Problems for Sect. 3.3

1. Show that the computational complexity of Algorithm 3.9 for solving DLP over $\mathbb{Z}_p^*$ is $\mathcal{O}((\log n)^{2+\epsilon})$, when $\log n$ is the number of bits of p.

2. The complexity of Algorithm 3.9 is currently in $\mathcal{BQP}$. Can this algorithm be improved to be in $\mathcal{QP}$? This is to say that can the randomness be removed from Algorithm 3.9?

3. In the general quantum DLP algorithm, the value of q is chosen to be in the range $p \leq q \leq 2p$. Can this value of q be reduced to a small number, so that the algorithm could be easy to implement on a small quantum computer?

4. Pollard's ρ and λ methods for DLP are very well suited for parallel computation, and in fact there are some novel parallel versions of the ρ and λ methods for DLP. Can the ρ and/or λ methods for DLP be implemented on a quantum computer? If so, develop a quantum version of the ρ or λ methods for DLP.

5. The NFS is currently the fastest method for solving DLP in $\mathbb{Z}_p^*$. Develop, if possible, a quantum version of the NFS for DLP.

6. The IFP and DLP can be generated to the HSP (hidden subgroup problem). Let G be an Abelian group. We say that $f : G \to S$ (taking values in some set S) hides the subgroup $H \leq G$ if

$$f(x) = f(y) \iff x - y \in H.$$

The Abelian HSP asks that given a device that computes f, find a generating set for H. Give a quantum algorithm to solve the more general HSP problem.

3.4 Conclusions, Notes, and Further Reading

Logarithms were invented by the Scottish mathematician John Napier (1550–1617). Basically, logarithm is the inverse of the mathematical operation exponentiation. We say k is the logarithm of y to the base x, denoted by $k = \log_x y$, if $y = x^k$, where $x, y, k \in \mathbb{R}$. The logarithm problem (LP) is to find k given x, y. Apparently, it is an easy problem, that is,

$$\text{LP}: \{x, y = x^k\} \xrightarrow{\text{easy}} \{k\},$$

as we can always solve the problem by using the following formulas:

$$\log_x y = \frac{\ln y}{\ln x}$$

and

$$\ln x = \sum_{i=1}^{\infty} (-1)^{i+1} \frac{(x-1)^i}{i}.$$

For example,

$$\log_2 5 = \frac{\ln 5}{\ln 2} \approx \frac{1.609437912}{0.692147106} \approx 2.321928095.$$

The situation is, however, completely different from that of DLP, say, e.g., over $\mathbb{Z}_p^*$ rather than over $\mathbb{R}$. Just the same as IFP, DLP is also an intractable computational number-theoretic problem and can be utilized to construct various public-key cryptosystems and protocols. There are many classical methods for solving DLP, say, for example:

1. Baby-step giant-step method

2. Pollard's ρ method

3. Pollard's λ method

4. Pohlig–Hellman method

5. Index calculus

6. Xedni calculus

7. Function field sieve (FFS)

It is interesting to note that for both IFP and DLP, no efficient algorithms are known for non-quantum computers, but efficient quantum algorithms are known. Moreover, algorithms from one problem are often adapted to the other, making IFP and DLP twin sister problems. In this chapter, we have introduced some of the most popular attacks on the DLP problem and some of the most widely used DLP-based cryptographic systems and protocols that are unbreakable by all classical attacks in polynomial time.

As mentioned, quantum computers can solve the DLP problem and break DLP-based cryptographic systems in polynomial time, so in the last section of this chapter, quantum attacks on DLP and DLP-based cryptography are discussed and analyzed.

The baby-step and giant-step method for DLP was originally proposed by Shanks in 1971 [52]. Pohlig–Hellman method for DLP was proposed in [40]. The ρ and λ methods for DLP were proposed by Pollard in [42]. The currently most powerful method, the index calculus, for DLP was discussed in many references such as [1, 21, 22, 49]. The FFS is based on the algebraic function field which is just an analog of the number field. Same as NFS, FFS can be used for solving both IFP and DLP. For more information on FFS, see [3, 4]. For general references on DLP and methods for solving DLP, readers are suggested to consult: [2, 5, 10, 12–14, 19, 23–26, 28–30, 33, 34, 37–39, 43–45, 47, 50, 58, 67, 69, 70].

DLP-based cryptography also forms an important class of cryptography, including cryptographic protocols and digital signatures. In the public literatures, the first public-key system, namely, the key-exchange scheme, was proposed by Diffie and Hellman in 1976 in [16], based on an idea of Merkle [35] (although published later). The first DLP-based cryptographic system and digital signature scheme were proposed by ElGamal in 1985 [19]. For general references on DLP-based cryptographic systems and digital signature schemes, readers are suggested to consult [1, 6–9, 15, 17, 18, 20, 24, 25, 27, 31, 34, 36, 40, 46, 48, 51, 59–67, 70], etc.

The quantum algorithm for DLP was first proposed in 1994 by Shor [53] (see Shor's other papers [54–57] for more information).

Molecular DNA-based computation for solving the DLP problem and for breaking DLP-based cryptography is a promising research area. Readers who are interested in this field are suggested to consult the paper by Chang et al. [11] and the references therein.

REFERENCES

[1]   L.M. Adleman, A subexponential algorithmic for the discrete logarithm problem with applications to cryptography, in *Proceedings of the 20th Annual IEEE Symposium on Foundations of Computer Science* (IEEE, New York, 1979), pp. 55–60

[2]   L.M. Adleman, Algorithmic number theory – the complexity contribution, in *Proceedings of the 35th Annual IEEE Symposium on Foundations of Computer Science* (IEEE, New York, 1994), pp. 88–113

[3]   L.M. Adleman, The function field Sieve, in *Algorithmic Number Theory (ANTS-I)*. Lecture Notes in Computer Science, vol. 877 (Springer, New York, 1994), pp. 108–121

[4]   L.M. Adleman, Function field Sieve method for discrete logarithms over finite fields. Inf. Comput. **151**, 5–16 (1999)

[5]   S. Bai, R.P. Brent, On the efficiency of Pollard's Rho method for discrete logarithms, in *Proceedings of the Fourteenth Computing: The Australasian Theory Symposium (CATS 2008)*, pp. 125–131, ed. by J. Harland, P. Manyem, Wollongong, NSW, Australia, 22–25 January 2008

[6]   T.H. Barr, *Invitation to Cryptology* (Prentice-Hall, Englewood Cliffs, 2002)

[7]   F.L. Bauer, *Decrypted Secrets – Methods and Maxims of Cryptology*, 3rd edn. (Springer, Berlin, 2002)

[8]   D. Bishop, *Introduction to Cryptography with Java Applets* (Jones and Bartlett, Burlington, MA, 2003)

[9]   J.A. Buchmann, *Introduction to Cryptography*, 2nd edn. (Springer, New York, 2004)

[10] J.A. Buchmann, D. Weber, Discrete logarithms: recent progress, in *Proceedings of an International Conference on Coding Theory, Cryptography and Related Areas*, ed. by J. Buchmann, T. Hoeholdt et al. (Springer, New York, 2000), pp. 42–56

[11] W.L. Chang, S.C. Huang, K.W. Lin, M.S.H. Ho, Fast parallel DNA-based algorithm for molecular computation: discrete logarithms. J. Supercomput. **56**(2), 129–163 (2011)

[12] H. Cohen, in *A Course in Computational Algebraic Number Theory*. Graduate Texts in Mathematics, vol. 138 (Springer, Berlin, 1993)

[13] H. Cohen, G. Frey, *Handbook of Elliptic and Hyperelliptic Curve Cryptography* (CRC Press, West Palm Beach, 2006)

[14] R. Crandall, C. Pomerance, *Prime Numbers – A Computational Perspective*, 2nd edn. (Springer, New York, 2005)

[15] W. Diffie, The first ten years of public-key cryptography. Proc. IEEE **76**(5), 560–577 (1988)

[16] W. Diffie, M.E. Hellman, New directions in cryptography. IEEE Trans. Inf. Theor. **22**(5), 644–654 (1976)

[17] W. Diffie, M.E. Hellman, Privacy and authentication: an introduction to cryptography. Proc. IEEE **67**(3), 397–427 (1979)

[18] A.J. Elbirt, *Understanding and Applying Cryptography and Data Security* (CRC Press, West Palm Beach, 2009)

[19] T. ElGamal, A public key cryptosystem and a signature scheme based on discrete logarithms. IEEE Trans. Inf. Theor. **31**, 469–472 (1985)

[20] B.A. Forouzan, *Cryptography and Network Security* (McGraw-Hill, New York, 2008)

[21] D.M. Gordon, Discrete logarithms in $GF(p)$ using the number field Sieve. SIAM J. Discrete Math. **6**(1), 124–138 (1993)

[22] D.M. Gordon, K.S. McCurley, Massively parallel computation of discrete logarithms, in *Advances in Cryptology - Crypto '92*. Lecture Notes in Computer Science, vol. 740 (Springer, New York, 1992), pp. 312–323

[23] T. Hayashi, N. Shinohara, L. Wang, S. Matsuo, M. Shirase, T. Takagi, Solving a 676-bit discrete logarithm problem in $GF(3^{6n})$, in *Public Key Cryptography - PKC 2010*. Lecture Notes in Computer Science, vol. 6056 (Springer, New York, 2010), pp. 351–367

[24] M.E. Hellman, An overview of public-key cryptography. IEEE Comm. Mag. 50th Anniversary Commemorative Issue **40**(5), 42–49 (1976, 2002)

[25] J. Hoffstein, J. Pipher, J.H. Silverman, *An Introduction to Mathematical Cryptography* (Springer, New York, 2008)

[26] M.D. Huang, W. Raskind, Signature calculus and discrete logarithm problems, in *ANTS 2006*. Lecture Notes in Computer Science, vol. 4076 (Springer, New York, 2006), pp. 558–572

[27] J. Katz, Y. Lindell, *Introduction to Modern Cryptography* (CRC Press, West Palm Beach, 2008)

[28] N. Koblitz, *A Course in Number Theory and Cryptography*, 2nd edn. Graduate Texts in Mathematics, vol. 114 (Springer, Berlin, 1994)

[29] N. Koblitz, in *Algebraic Aspects of Cryptography*. Algorithms and Computation in Mathematics, vol. 3 (Springer, New York, 1998)

[30] M.T. Lacey, *Cryptography, Cards, and Kangaroos* (Georgia Institute of Technology, Atlanta, 2008)

[31] W. Mao, *Modern Cryptography* (Prentice-Hall, Englewood Cliffs, 2004)

[32] J.L. Massey, J.K. Omura, *Method and Apparatus for Maintaining the Privacy of Digital Message Conveyed by Public Transmission*, US Patent No 4677600, 28 Jan 1986

[33] K.S. McCurley, The discrete logarithm problem, in *Cryptology and Computational Number Theory*, ed. by C. Pomerance. Proceedings of Symposia in Applied Mathematics, vol. 42 (American Mathematics Society, Providence, 1990), pp. 49–74

[34] A. Menezes, P.C. van Oorschot, S.A. Vanstone, *Handbook of Applied Cryptosystems* (CRC Press, West Palm Beach, 1996)

[35] R.C. Merkle, Secure Communications over insecure channels. Comm. ACM **21**, 294–299 (1978) (submitted in 1975)

[36] R.A. Mollin, *An Introduction to Cryptography*, 2nd edn. (Chapman & Hall/CRC Press, London/West Palm Beach, 2006)

[37] R. Motwani, P. Raghavan, *Randomized Algorithms* (Cambridge University Press, Cambridge, 1995)

[38] A.M. Odlyzko, Discrete logarithms in finite fields and their cryptographic significance, in *Advances in Cryptography, EUROCRYPT '84*. Proceedings, Lecture Notes in Computer Science, vol. 209 (Springer, Berlin, 1984), pp. 225–314

[39] A.M. Odlyzko, Discrete logarithms: the past and the future. Des. Codes Cryptography **19**, 129–145 (2000)

[40] S.C. Pohlig, M. Hellman, An improved algorithm for computing logarithms over GF(p) and its cryptographic significance. IEEE Trans. Inf. Theor. **24**, 106–110 (1978)

[41] J.M. Pollard, A Monte Carlo method for factorization. BIT **15**, 331–332 (1975)

[42] J.M. Pollard, Monte Carlo methods for index computation (mod p). Math. Comput. **32**, 918–924 (1980)

[43] J.M. Pollard, Kangaroos, monopoly and discrete logarithms. J. Cryptol. **13**, 437–447 (2000)

[44] J.M. Pollard, Kruskal's card trick. Math. Gazette **84**, 500, 265–267 (2000)

[45] C. Pomerance, Elementary thoughts on discrete logarithms, in *Algorithmic Number Theory*, ed. by J.P. Buhler, P. Stevenhagen (Cambridge University Press, Cambridge, 2008), pp. 385–395

[46] M. Rabin, Digitalized Signatures and Public-Key Functions as Intractable as Factorization. Technical Report MIT/LCS/TR-212, MIT Laboratory for Computer Science (1979)

[47] H. Riesel, *Prime Numbers and Computer Methods for Factorization* (Birkhäuser, Boston, 1990)

[48] J. Rothe, *Complexity Theory and Cryptography* (Springer, New York, 2005)

[49] O. Schirokauer, D. Weber, T. Denny, Discrete logarithms: the effectiveness of the index calculus method, in *Algorithmic Number Theory (ANTS-II)*. Lecture Notes in Computer Science, vol. 1122 (Springer, New York, 1996), pp. 337–362

[50] O. Schirokauere, The impact of the number field Sieve on the discrete logarithm problem in finite fields, in *Algorithmic Number Theory*, ed. by J.P. Buhler, P. Stevenhagen (Cambridge University Press, Cambridge, 2008), pp. 421–446

[51] B. Schneier, *Applied Cryptography: Protocols, Algorithms, and Source Code in C*, 2nd edn. (Wiley, New York, 1996)

[52] D. Shanks, Class number, a theory of factorization and Genera, in *Proceedings of Symposium of Pure Mathematics*, vol. 20 (AMS, Providence, 1971), pp. 415–440

[53] P. Shor, Algorithms for quantum computation: discrete logarithms and factoring, in *Proceedings of the 35th Annual Symposium on Foundations of Computer Science*, Santa Fe, NM, 20–22 November (IEEE Computer Society, Silver Spring, 1994), pp. 124–134

[54] P. Shor, Polynomial-time algorithms for prime factorization and discrete logarithms on a quantum computer. SIAM J. Comput. **26**(5), 1484–1509 (1997)

[55] P. Shor, Quantum computing. Documenta Math. Extra Volume ICM **1**, 467–486 (1998)

[56] P. Shor, Polynomial-time algorithms for prime factorization and discrete logarithms on a quantum computer. SIAM Rev. **41**(2), 303–332 (1999)

[57] P. Shor, Introduction to quantum algorithms. AMS Proc. Symp. Appl. Math. **58**, 17 (2002)

[58] V. Shoup, *A Computational Introduction to Number Theory and Algebra* (Cambridge University Press, Cambridge, 2005)

[59] N. Smart, *Cryptography: An Introduction* (McGraw-Hill, New York, 2003)

[60] M. Stamp, R.M. Low, *Applied Cryptanalysis* (Wiley, New York, 2007)

[61] A. Stanoyevitch, *Introduction to Cryptography* (CRC Press, West Palm Beach, 2011)

[62] D.R. Stinson, *Cryptography: Theory and Practice*, 3rd edn. (Chapman & Hall/CRC Press, London/West Palm Beach, 2006)

[63] C. Swenson, *Modern Cryptanalysis* (Wiley, New York, 2008)

[64] The digital signature standard proposed by NIST and responses to NIST's proposal. Comm. ACM **35**(7), 36–54 (1992)

[65] W. Trappe, L. Washington, *Introduction to Cryptography with Coding Theory*, 2nd edn. (Prentice-Hall, Englewood Cliffs, 2006)

[66] H.C.A. van Tilborg, *Fundamentals of Cryptography* (Kluwer, Dordrecht, 1999)

[67] S.S. Wagstaff Jr., *Cryptanalysis of Number Theoretic Ciphers* (Chapman & Hall/CRC Press, London/West Palm Beach, 2002)

[68] D. Weber, T.F. Denny, The solution of McCurley's discrete log challenge, in *Advances in Cryptology - CRYPTO '98*. Lecture Notes in Computer Science, vol. 1462 (Springer, Berlin, 1998), pp. 458–471

[69] S.Y. Yan, Computing prime factorization and discrete logarithms: from index calculus to Xedni calculus. Int. J. Comput. Math. **80**(5), 573–590 (2003)

[70] S.Y. Yan, in *Primality Testing and Integer Factorization in Public-Key Cryptography*. Advances in Information Security, vol. 11, 2nd edn. (Springer, New York, 2009)

4. Quantum Attacks on ECDLP-Based Cryptosystems

> *The existing scientific concepts cover always only a very limited part of reality, and the other part that has not yet been understood is infinite.*
>
> WERNER HEISENBERG (1901–1976)
> The 1932 Nobel Laureate in Physics

In this chapter we shall first study the elliptic curve discrete logarithm problem (ECDLP) and the classical solutions to ECDLP, and then we shall discuss some of the most popular ECDLP-based cryptographic systems for which there is no efficient cryptanalytic algorithm. Finally, we shall introduce some quantum algorithms for attacking both the ECDLP problem and the ECDLP-based cryptographic systems.

4.1 ECDLP and Classical Solutions

Definition of ECDLP

The ECDLP: Let E be an elliptic curve over the finite field $\mathbb{F}_p$, say, given by a Weierstrass equation

$$E : \ y^2 \equiv x^3 + ax + b \ (\mathrm{mod} \ p), \qquad (4.1)$$

S and T the two points in the elliptic curve group $E(\mathbb{F}_p)$. Then the ECDLP is to find the integer k (assuming that such an integer k exists)

$$k = \log_T S \in \mathbb{Z} \quad \text{or} \quad k \equiv \log_T S \ (\mathrm{mod} \ p) \qquad (4.2)$$

such that

$$S = kT \in E(\mathbb{F}_p) \quad \text{or} \quad S \equiv kT \ (\mathrm{mod} \ p). \qquad (4.3)$$

S.Y. Yan, *Quantum Attacks on Public-Key Cryptosystems*,
DOI 10.1007/978-1-4419-7722-9_4,
© Springer Science+Business Media, LLC 2013

The ECDLP is a more difficult problem than the DLP, on which the elliptic curve digital signature algorithm (ECDSA) is based on. Clearly, the ECDLP is the generalization of DLP, which extends the multiplicative group $\mathbb{F}_p^*$ to the elliptic curve group $E(\mathbb{F}_p)$.

Pohlig–Hellman for ECDLP

The ECDLP problem is a little bit more difficult than the DLP problem, on which the elliptic curve digital signature algorithm/elliptic curve digital signature standard (ECDSA/ECDSS) [26] is based on. As ECDLP is the generalization of DLP, which extends, e.g., the multiplicative group $\mathbb{F}_p^*$ to the elliptic curve group $E(\mathbb{F}_p)$, many methods for DLP, even for IFP, can be extended to ECDLP, for example, the baby-step giant-step for DLP and Pollard's ρ and λ methods for IFP and DLP; Silver–Pohlig–Hellman method for DLP can also be naturally extended to ECDLP. In what follows, we present an example of solving ECDLP by an analog of Silver–Pohlig–Hellman method for elliptic curves over $\mathbb{F}_p^*$.

Example 4.1. Let
$$Q \equiv kP \pmod{1009}$$
where
$$\begin{cases} E: \quad y^2 \equiv x^3 + 71x + 602 \pmod{1009} \\ P = (1, 237) \\ Q = (190, 271) \\ \text{order}(E(\mathbb{F}_{1009})) = 1060 = 2^2 \cdot 5 \cdot 53 \\ \text{order}(P) = 530 = 2 \cdot 5 \cdot 53 \end{cases}$$
Find k.

[1] Find the individual logarithm modulo 2: as $(530/2) = 265$, we have
$$\begin{cases} P_2 = 265P = (50, 0) \\ Q_2 = 265Q = (50, 0) \\ Q_2 = P_2 \\ k \equiv 1 \pmod{2} \end{cases}$$

[2] Find the individual logarithm modulo 5: as $530/5 = 106$, we have
$$\begin{cases} P_5 = 106P = (639, 160) \\ Q_5 = 106Q = (639, 849) \\ Q_5 = -P_5 \\ k \equiv 4 \pmod{5} \end{cases}$$

[3] Find the individual logarithm modulo 53: as $530/53 = 10$, we have

$$\begin{cases} P_{53} = 10P = (32, 737) \\ Q_{53} = 10Q = (592, 97) \\ Q_{53} = 48P_{53} \\ k \equiv 48 \pmod{53} \end{cases}$$

[4] Use the Chinese Remainder Theorem to combine the individual logarithms to get the final logarithm:

$$\mathrm{CHREM}([1, 4, 48], [2, 5, 53]) = 419.$$

That is,

$$(190, 271) \equiv 419(1, 237) \pmod{1009}$$

or, alternatively,

$$(190, 271) \equiv \underbrace{(1, 237) + \cdots + (1, 237)}_{419 \text{ summands}} \pmod{1009}.$$

Baby-Step Giant-Step for ECDLP

The Shank's baby-step giant-step for DLP can be easily extended for ECDLP. To find k in $Q = kP$, the idea is to compute and store a list of points iP for $1 \leqslant i \leqslant m$ (baby steps) and then compute $Q - jmP$ (giant steps) and try to find a match in the stored list. The algorithm may be described as follows:

Algorithm 4.1 (Baby-Step Giant-Step for ECDLP). Let E be an elliptic curve over $\mathbb{Z}_n$, $P, Q \in E(\mathbb{Z}_n)$. This algorithm tries to find k in $Q \equiv kP \pmod{n}$.

[1] Set $m = \lfloor n \rfloor$.

[2] For i from 1 to m, compute and store iP.

[3] For j from 1 to $m - 1$, compute $Q - jmP$ and check this against the list stored in Step [2].

[4] If a match is found then $Q - jmP = iP$ and hence $Q = (i + jm)P$.

[5] Output $k \equiv i + jm \pmod{n}$.

Example 4.2 (Baby-Step Giant-Step for ECDLP). Let $E \backslash \mathbb{F}_{727} : y^2 \equiv x^3 + 231x + 508 \pmod{719}$ be an elliptic curve over $\mathbb{F}_{719}$, $|E(\mathbb{F}_{719})| = 727$, $P = (513, 30)$, $Q = (519, 681) \in E(\mathbb{F}_{719})$. We wish to find k $Q \equiv kP \pmod{719}$.

[1] Set $m = \lfloor 719 \rfloor = 27$ and compute $27P = (714, 469)$.

[2] For i from 1 to m, compute and store iP:

$$1P = (513, 30)$$
$$2P = (210, 538)$$
$$3P = (525, 236)$$
$$4P = (507, 58)$$
$$5P = (427, 421)$$
$$6P = (543, 327)$$
$$\vdots$$
$$24P = (487, 606)$$
$$25P = (529, 253)$$
$$26P = (239, 462)$$
$$27P = (714, 469).$$

[3] For j from 1 to $m - 1$, compute $Q - jmP$ and check this against the list stored in Step [2].

$$Q - (0 \cdot 27)P = (511, 681)$$
$$Q - (1 \cdot 27)P = (650, 450)$$
$$Q - (2 \cdot 27)P = (95, 422)$$
$$\vdots$$
$$Q - (19 \cdot 27)P = (620, 407)$$
$$Q - (20 \cdot 27)P = (143, 655)$$
$$Q - (21 \cdot 27)P = (239, 462).$$

[4] A match is found for $27P = (714, 469)$ and $Q - (21 \cdot 27)P = (239, 462)$. Thus, $Q = (26 + 21 \cdot 27)P$.

[5] Output $k \equiv 26 + 21 \cdot 27 \equiv 593 \pmod{727}$.

ρ Method for ECDLP

The fastest algorithm for solving ECDLP is Pollard's ρ method. Up to date, the largest ECDLP instance solved with ρ is still the ECC_p-109, for an elliptic curve over a 109-bit prime field. Recall that the ECDLP problem asks to find $k \in [1, r - 1]$ such that

$$Q = kP,$$

where r is a prime number, P is a point of order r on an elliptic curve over a finite field $\mathbb{F}_p$, $Q \in G$, and $G = \langle P \rangle$. The main idea of ρ for ECDLP is to find distinct pairs (c', d') and (c'', d'') of integers modulo r such that

$$c'P + d'Q = c''P + d''Q.$$

Then
$$(c' - c'')P = (d'' - d')Q,$$
that is,
$$Q = \frac{c' - c''}{d'' - d'}P,$$
thus,
$$k \equiv \frac{c' - c''}{d'' - d'} \pmod{r}.$$

To implement the idea, we first choose a random iteration function $f : G \to G$ and then start a random initial point P_0 and compute the iterations $P_{i+1} = f(P_i)$. Since G is finite, there will be some indices $i < j$ such that $P_i = P_j$. Then
$$P_{i+1} = f(P_i) = f(P_j) = P_{j+1},$$
and in fact
$$P_{i+l} = P_{j+l}, \text{ for all } l \geqslant 0.$$

Therefore, the sequence of points $\{P_i\}$ is periodic with period $j - i$ (see Fig. 4.1). This is why we call it the ρ method; we may also call it the λ method, as the computation paths for $c'P + d'Q$ and $c''P + d''Q$ will eventually be met and traveled along on the same road, symbolized by the Greek letter λ. If f is a randomly chosen random function, then we expect to find a match (i.e., a collision) with j at most a constant times $\sqrt{r}$. In fact, by the birthday paradox, the expected number of iterations before a *collision* is obtained is approximately $\sqrt{\pi r/2} \approx 1.2533\sqrt{r}$. To quickly detect the collision, the Floyd

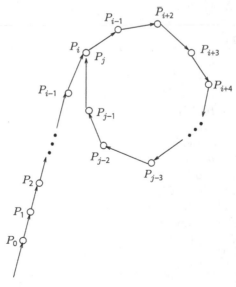

Figure 4.1. ρ for ECDLP

cycle detection trick will be used. That is, just the same as ρ for IFP and DLP, we compute pairs (P_i, P_{2i}) for $i = 1, 2' \cdots$, until a match is found. Here is the algorithm and an example [18].

Algorithm 4.2 (Pollard's ρ Algorithm for ECDLP). Given $P \in E(\mathbb{F}_q)$ of prime order r, $Q \in \langle P \rangle$, this algorithm tries to find

$$k \equiv \log_P Q \ (\text{mod } n)$$

such that

$$Q \equiv kP \ (\text{mod } p),$$

via

$$k \equiv \frac{c' - c''}{d'' - d'} \ (\text{mod } r).$$

[1] Initialization. Choose the number L of branches, and select a partition function $H : \langle P \rangle \rightarrow \{1, 2, \ldots, L\}$.

[2] Compute $a_i P + b_i Q$.

> for i from 1 to L do
>   choose $a_i, b_i \in [0, r-1]$
>   compute $R_i = a_i P + b_i Q$.

[3] Compute $c'P + d'Q$. Choose $c', d' \in [0, r-1]$, and compute $X' = c'P + d'Q$.

[4] Prepare for loop.

> Set $X'' \leftarrow X'$
> $c'' \leftarrow c'$
> $d'' \leftarrow d'$.

[5] Loop.

> Repeat
> Compute $j = H(X')$
> Set $X' \leftarrow X' + R_j$
> $c' \leftarrow c' + a_j \bmod r$
> $d' \leftarrow d' + b_j \bmod r$.
> for i from to 2 do
>   Compute $j = H(X'')$
> Set $X'' \leftarrow X'' + R_j$
> $c'' \leftarrow c'' + a_j \bmod r$
> $d'' \leftarrow d'' + b_j \bmod r$.
> Until $X' = X''$.

[6] Output and exit.

If $d \neq d''$ then computer $k \equiv (c' - c'')(d'' - d')^{-1} \pmod{r}$.
otherwise return(failure), stop or startover again.

Example 4.3. Consider the elliptic curve

$$E \backslash \mathbb{F}_{229} : y^2 \equiv x^3 + x + 44 \pmod{229}.$$

The point $P = (5, 116) \in E(\mathbb{F}_{229})$ has prime order $r = 239$. Let $Q = (155, 166) \in \langle P \rangle$ (where $\langle P \rangle$ denotes the subgroup generated by the point P). We wish to find k such that

$$Q \equiv kP \pmod{229}.$$

That is,

$$k \equiv \log_P Q \pmod{239}.$$

We perform the following steps:

[1] Select the partition function $H : \langle P \rangle \to \{1, 2, 3, 4\}$ with 4 partitions:

$$H(x, y) = (x \bmod 4) = 1.$$

Let $R_i = a_i P + b_i Q$ with $i = 1, 2, 3, 4$. Then

$(a_1, b_1, R_1) = (79, 163, (135, 117))$
$(a_2, b_2, R_2) = (206, 19, (96, 97))$
$(a_3, b_3, R_3) = (87, 109, (84, 62))$
$(a_4, b_4, R_4) = (219, 68, (72, 134))$.

[2] Compute the iteration table until a mach (collision) is found.

Iteration	c'	d'	$c'P + d'Q$	c''	d''	$c''P + d''Q$
0	54	175	(39,159)	54	175	(39,159)
1	34	4	(160,9)	113	167	(130,182)
2	113	167	(130,182)	180	105	(36, 97)
3	200	37	(27,17)	0	97	(108,89)
4	180	105	(36,97)	46	40	(223,153)
5	20	29	(119,180)	232	127	(167,57)
6	0	97	(108,89)	192	24	(57,105)
7	79	21	(81,168)	139	111	(185,227)
8	46	40	(223,153)	193	0	(197,92)
9	26	108	(9,18)	140	87	(194,145)
10	232	127	(167,57)	67	120	(223,153)
11	212	195	(75,136)	14	207	(167,57)
12	192	24	**(57,105)**	213	104	**(57,105)**

[3] At the step $i = 12$, we find a match

$$192P + 24Q = 213P + 104Q = (\mathbf{57}, \mathbf{105}).$$

That is,

$$Q = \frac{192 - 213}{104 - 24}P \pmod{229}.$$

Thus, we have

$$
\begin{aligned}
k &\equiv (192 - 213)(104 - 24)^{-1} \\
  &\equiv 176 \pmod{239}.
\end{aligned}
$$

Xedni Calculus for ECDLP

The index calculus is the most powerful method for DLP in some groups including the multiplicative group $\mathbb{F}_q^*$ over a finite field; it is however generally not suitable for ECDLP as it is not for general groups. In what follows, we introduce a method called xedni calculus for ECDLP.

The xedni calculus was first proposed by Joseph Silverman in 1998 [55] and analyzed in [24, 34, 57]. It is called *xedni calculus* because it "stands index calculus on its head." The xedni calculus is a new method that *might* be used to solve the ECDLP, although it has not yet been tested in practice. It can be described as follows [55]:

[1] Choose points in $E(\mathbb{F}_p)$ and lift them to points in $\mathbb{Z}^2$.

[2] Choose a curve $E(\mathbb{Q})$ containing the lift points; use Mestre's method [44] (in reverse) to make rank $E(\mathbb{Q})$ small.

Whilst the index calculus works in reverse:

[1] Lift $E/\mathbb{F}_p$ to $E(\mathbb{Q})$; use Mestre's method to make rank $E(\mathbb{Q})$ large.

[2] Choose points in $E(\mathbb{F}_p)$ and try to lift them to points in $E(\mathbb{Q})$.

A brief description of the xedni algorithm is as follows (a complete description and justification of the algorithm can be found in [55]).

Algorithm 4.3 (Xedni calculus for the ECDLP). Let $\mathbb{F}_p$ be a finite field with p elements (p prime); $E/\mathbb{F}_p$ an elliptic curve over $\mathbb{F}_p$, say, given by

$$E: \quad y^2 + a_{p,1}xy + a_{p,3}y = x^3 + a_{p,2}x^2 + a_{p,4}x + a_{p,6};$$

N_p the number of points in $E(\mathbb{F}_p)$; and S and T the two points in $E(\mathbb{F}_p)$. This algorithm tries to find an integer k

$$k = \log_T S$$

such that

$$S = kT \quad \text{in } E(\mathbb{F}_p).$$

[1] Fix an integer $4 \leqslant r \leqslant 9$ and an integer M which is a product of small primes.

[2] Choose r points:

$$P_{M,i} = [x_{M,i}, y_{M,i}, z_{M,i}], \quad 1 \leqslant i \leqslant r \qquad (4.4)$$

having integer coefficients and satisfying
[a] the first 4 points are $[1,0,0]$, $[0,1,0]$, $[0,0,1]$, and $[1,1,1]$.

[b] For every prime $l \mid M$, the matrix $\mathbf{B}(P_{M,1}, \ldots, P_{M,r})$ has maximal rank modulo l.
Further choose coefficients $u_{M,1}, \ldots, u_{M,10}$ such that the points $P_{M,1}, \ldots, P_{M,r}$ satisfy the congruence:

$$u_{M,1}x^3 + u_{M,2}x^2y + u_{M,3}xy^2 + u_{M,4}y^3 + u_{M,5}x^2z + u_{M,6}xyz + u_{M,7}y^2z$$
$$+u_{M,8}xz^2 + u_{M,9}yz^2 + u_{M,10}z^3 \equiv 0 \pmod{M}. \qquad (4.5)$$

[3] Choose r random pair of integers (s_i, t_i) satisfying $1 \leqslant s_i, t_i < N_p$, and for each $1 \leqslant i \leqslant r$, compute the point $P_{p,i} = (x_{p,i}, y_{p,i})$ defined by

$$P_{p,i} = s_i S - t_i T \quad \text{in } E(\mathbb{F}_p). \qquad (4.6)$$

[4] Make a change of variables in $\mathbb{P}^2$ of the form

$$\begin{pmatrix} X' \\ Y' \\ Z' \end{pmatrix} = \begin{pmatrix} a_{11} & a_{12} & a_{13} \\ a_{21} & a_{22} & a_{23} \\ a_{31} & a_{32} & a_{33} \end{pmatrix} \begin{pmatrix} X \\ Y \\ Z \end{pmatrix} \qquad (4.7)$$

so that the first four points become

$$P_{p,1} = [1,0,0], \ P_{p,2} = [0,1,0], \ P_{p,3} = [0,0,1], \ P_{p,4} = [1,1,1].$$

The equation for E will then have the form:

$$u_{p,1}x^3 + u_{p,2}x^2y + u_{p,3}xy^2 + u_{p,4}y^3 + u_{p,5}x^2z + u_{p,6}xyz$$
$$+u_{p,7}y^2z + u_{p,8}xz^2 + u_{p,9}yz^2 + u_{p,10}z^3 = 0. \qquad (4.8)$$

[5] Use the Chinese Remainder Theorem to find integers $u'_1, \ldots, u'_{10}$ satisfying

$$u'_i \equiv u_{p,i} \pmod{p} \text{ and } u'_i \equiv u_{M,i} \pmod{M} \text{ for all } 1 \leqslant i \leqslant 10. \qquad (4.9)$$

[6] Lift the chosen points to $\mathbb{P}^2(\mathbb{Q})$. That is, choose points

$$P_i = [x_i, y_i, z_i], \quad 1 \leqslant i \leqslant r, \tag{4.10}$$

with integer coordinates satisfying

$$P_i \equiv P_{p,i} \pmod{p} \text{ and } P_i \equiv P_{M,i} \pmod{M} \text{ for all } 1 \leqslant i \leqslant r. \tag{4.11}$$

In particular, take $P_1 = [1, 0, 0], P_2 = [0, 1, 0], P_3 = [0, 0, 1], P_4 = [1, 1, 1]$.

[7] Let $\mathbf{B} = \mathbf{B}(P_1, \ldots, P_r)$ be the matrix of cubic monomials defined earlier. Consider the system of linear equations:

$$\mathbf{B}\mathbf{u} = 0. \tag{4.12}$$

Find a small integer solution $\mathbf{u} = [u_1, \ldots, u_{10}]$ to (4.12) which has the additional property

$$\mathbf{u} \equiv [u_1', \ldots, u_{10}'] \pmod{M_p}, \tag{4.13}$$

where $u_1', \ldots, u_{10}'$ are the coefficients computed in Step [5]. Let $C_{\mathbf{u}}$ denote the associated cubic curve:

$$C_{\mathbf{u}}: \ u_1 x^3 + u_2 x^2 y + u_3 xy^2 + u_4 y^3 + u_5 x^2 z + u_6 xyz$$
$$+u_7 y^2 z + u_8 xz^2 + u_9 yz^2 + u_{10} z^3 = 0. \tag{4.14}$$

[8] Make a change of coordinates to put $C_{\mathbf{u}}$ into standard minimal Weierstrass form with the point $P_1 = [1, 0, 0]$, the point at infinity, $\mathcal{O}$. Write the resulting equation as

$$E_{\mathbf{u}}: \ y^2 + a_1 xy + a_3 y = x^3 + a_2 x^2 + a_4 x + a_6 \tag{4.15}$$

with $a_1, \ldots, a_6 \in \mathbb{Z}$, and let $Q_1, Q_2, \ldots, Q_r$ denote the images of $P_1, P_2, \ldots,$ P_r under this change of coordinates (so in particular, $Q_1 = \mathcal{O}$). Let $c_4(\mathbf{u})$, $c_6(\mathbf{u})$, and $\Delta(\mathbf{u})$ be the usual quantities in [55] associated to (4.15).

[9] Check if the points $Q_1, Q_2, \ldots, Q_r \in E_{\mathbf{u}}(\mathbb{Q})$ are independent. If they are, return to Step [2] or [3]. Otherwise compute a relation of dependence

$$n_2 Q_2 + n_3 Q_3 + \cdots + n_r Q_r = \mathcal{O}, \tag{4.16}$$

set

$$n_1 = -n_2 - n_3 - \cdots - n_r, \tag{4.17}$$

and continue with the next step.

[10] Compute

$$s = \sum_{i=1}^{r} n_i s_i \quad \text{and} \quad t = \sum_{i=1}^{r} n_i t_i. \qquad (4.18)$$

If $\gcd(s, n_p) > 1$, go to Step [2] or [3]. Otherwise compute an inverse $ss' \equiv 1 \pmod{N_p}$. Then

$$\log_T S \equiv s't \pmod{N_p}, \qquad (4.19)$$

and the ECDLP is solved.

As can be seen, the basic idea in the above algorithm is that we first choose points $P_1, P_2, \ldots, P_r$ in $E(\mathbb{F}_p)$ and lift them to points $Q_1, Q_2, \ldots, Q_r$ having integer coordinates, and then we choose an elliptic curve $E(\mathbb{Q})$ that goes through the points $Q_1, Q_2, \ldots, Q_r$; finally, check if the points $Q_1, Q_2, \ldots, Q_r$ are *dependent*. If they are, the ECDLP is almost solved. Thus, the goal of the xedni calculus is to find an instance where an elliptic curve has *smaller* than expected rank. Unfortunately, a set of points $Q_1, Q_2, \ldots, Q_r$ as constructed above will usually be *independent*. So, it will not work. To make it work, a congruence method, due to Mestre [44], is used *in reverse* to produce the lifted curve E having smaller than expected rank.[1] Again unfortunately, Mestre's method is based on some deep ideas and unproved conjectures in analytic number theory and arithmetic algebraic geometry; it is not possible for us at present to give even a rough estimate of the algorithm's running time. So, virtually we know nothing about the complexity of the xedni calculus. We also do not know if the xedni calculus will be practically useful; it may be completely useless from a practical point of view. Much needs to be done before we can have a better understanding of the xedni calculus.

The index calculus is probabilistic, subexponential-time algorithm applicable for both the integer factorization problem (IFP) and the finite field discrete logarithm problem (DLP). However, there is no known subexponential-time algorithm for the ECDLP; the index calculus will not work for the ECDLP. The *xedni calculus*, on the other hand, is applicable to ECDLP (it is in fact also applicable to IFP and DLP), but unfortunately its complexity is essentially unknown. From a computability point of view, xedni calculus is applicable to IFP, DLP, and ECDLP, but from a complexity point of view, the xedni calculus may turn out to be useless (i.e., not at all practical). As for quantum algorithms, we now know that IFP, DLP, and ECDLP can all be solved in polynomial time if a quantum computer is available for use. However, the problem with quantum algorithms is that a practical quantum computer is out of reach in today's technology. We summarize various algorithms for IFP, DLP, and ECDLP in Table 4.1.

Finally, we conclude that we do have algorithms to solve the IFP, DLP, and ECDLP; the only problem is that we do not have an efficient algorithm

---

[1]Mestre's original method is to produce elliptic curves of large rank.

Table 4.1. Algorithms for IFP, DLP, and ECDLP

IFP	DLP	ECDLP
Trial divisions		
	Baby-step giant-step	Baby-step giant-step
	Pohlig–Hellman	Pohlig–Hellman
ρ	ρ	ρ
CFRAC/MPQS	Index calculus	
NFS	NFS	
Xedni calculus	Xedni calculus	Xedni calculus
Quantum algorithm	Quantum algorithms	Quantum algorithms

nor does anyone proved that no such an efficient algorithm exists. From a computational complexity point of view, a $\mathcal{P}$-type problem is easy to solve, whereas an $\mathcal{NP}$-type problem is easy to verify [17], so the IFP, DLP, and ECDLP are clearly in $\mathcal{NP}$. For example, it might be difficult (indeed, it is difficult at present) to factor a large integer, but it is easy to verify whether or not a given factorization is correct. If $\mathcal{P} = \mathcal{NP}$, then the two types of the problems are the same, the factorization is difficult only because no one has been clever enough to find an easy/efficient algorithm yet (it may turn out that the IFP is indeed $\mathcal{NP}$-hard, regardless of the cleverness of the human beings). Whether or not $\mathcal{P} = \mathcal{NP}$ is one of the biggest open problems in both mathematics and computer science, and it is listed in the first of the seven Millennium Prize Problems by the Clay Mathematics Institute in Boston on 24 May 2000 [11]. The struggle continues and more research needs to be done before we can say anything about whether or not $\mathcal{P} = \mathcal{NP}$!

Exercises and Problems for Sect. 4.1

1. As Shank's baby-step giant-step method works for arbitrary groups, it can be extended, of course, to elliptic curve groups.
   (a) Develop an elliptic curve analog of Shank's algorithm to solve the ECDLP problem.

   (b) Use the analog algorithm to solve the following ECDLP problem, that is, to find k such that

   $$Q \equiv kP \pmod{41}$$

   where $E/\mathbb{F}_{41} : y^2 \equiv x^3 + 2x + 1 \pmod{41}$, $P = (0,1)$, and $Q = (30, 40)$.

2. Poland's ρ and λ methods for IFP/DLP can also be extended to ECDLP.
   (a) Develop an elliptic curve analog of Poland ρ algorithm to solve the ECDLP problem.

(b) Use the ρ algorithm to solve the following ECDLP problem: find k such that

$$Q \equiv kP \pmod{p}$$

where $E\backslash\mathbb{F}_{1093} : y^2 \equiv x^3 + x + 1 \pmod{1093}$, $P = (0, 1)$, and $Q = (413, 959)$.

3. (Extend the Silver–Pohlig–Hellman method)

   (a) Develop an elliptic curve analog of Silver–Pohlig–Hellman method for ECDLP.

   (b) Use this analog method to solve the following ECDLP problem: find k such that

   $$Q \equiv kP \pmod{p}$$

   where $E\backslash\mathbb{F}_{599} : y^2 \equiv x^3 + 1 \pmod{1093}$, $P = (60, 19)$, and $Q = (277, 239)$.

4. In 1993, Menezes, Okamoto, and Vanstone developed an algorithm for ECDLP over $\mathbb{F}_{p^m}$ with p^m prime power. Give a description and complexity analysis of this algorithm.

5. Let $E\backslash\mathbb{F}_p$ be the elliptic curve E over $\mathbb{F}_p$ with p prime, where E is defined by

$$y^2 = x^3 + ax + b.$$

   (a) Let $P, Q \in E$ with $P \neq \pm Q$ are two points on E. Find the addition formula for computing $P + Q$.

   (b) Let $P \in E$ with $P \neq -P$. Find the addition formula for computing $2P$.

   (c) Let $E\backslash\mathbb{F}_{23}$ be as follows:

   $$E\backslash\mathbb{F}_{23} : y^2 \equiv x^3 + x + 4 \pmod{23}.$$

   Find all the points, $E(\mathbb{F}_{23})$, including the point at infinity, on the E.

   (d) Let $P = (7, 20)$ and $Q = (17, 14)$ be in $E\backslash\mathbb{F}_{23}$ defined above, and find $P + Q$ and $2P$.

   (e) Let $Q = (13, 11)$ and $P = (0, 2)$ such that $Q \equiv kP \pmod{23}$. Find $k = \log_P Q \pmod{23}$, the discrete logarithm over $E(\mathbb{F}_{23})$.

6. Let the elliptic curve be as follows:

$$E\backslash\mathbb{F}_{151} : y^2 \equiv x^3 + 2x \pmod{151}$$

   with order 152. A point $P = (97, 26)$ with order 19 is given. Let also $Q = (43, 4)$ such that

$$Q \equiv kP \pmod{151}.$$

Find $k = \log_P Q \pmod{151}$, the discrete logarithm over $E(\mathbb{F}_{151})$.

7. Let the elliptic curve be as follows:

$$E \backslash \mathbb{F}_{43} : y^2 \equiv x^3 + 39x^2 + x + 41 \pmod{43}$$

with order 43. Find the ECDLP

$$k = \log_P Q \pmod{43},$$

where $P = (0, 16)$ and $Q = (42, 32)$.

8. Let the elliptic curve be as follows:

$$E \backslash \mathbb{F}_{1009} : y^2 \equiv x^3 + 71x + 602 \pmod{1009}.$$

Find the ECDLP

$$k' = \log'_P Q' \pmod{53}$$

in

$$Q' = (529, 97) = k'(32, 737) = k'P'$$

in the subgroup of order 53 generated by $P' = (32, 737)$.

9. In ECCp-109, given

$$E \backslash \mathbb{F}_p : \ y^2 \equiv x^3 + ax + b \pmod{p},$$
$$\{P(x_1, y_1), Q(x_2, y_2)\} \in E(\mathbb{F}_p)$$
$$p = 564538252084441556247016902735257,$$
$$a = 321094768129147601892514872825668,$$
$$b = 430782315140218274262276694323197,$$
$$x_1 = 97339010987059066523156133908935,$$
$$y_1 = 149670372846169285760682371978898,$$
$$x_2 = 44646769697405861057630861884284,$$
$$y_2 = 522968098895785888047540374779097,$$

show that the following value of k

$$k = 281183840311601949668207954530684$$

is the correct value satisfying

$$Q(x_2, y_2) \equiv k \cdot P(x_1, y_1) \pmod{p}.$$

10. In ECCp-121, given

$$E \backslash \mathbb{F}_p : \ y^2 \equiv x^3 + ax + b \ (\text{mod } p),$$
$$\{P(x_1, y_1), Q(x_2, y_2)\} \in E(\mathbb{F}_p),$$
$$p = 4451685225093714772084598273548427,$$
$$a = 4451685225093714772084598273548424,$$
$$b = 2061118396808653202902996166388514,$$
$$x_1 = 188281465057972534892223778713752,$$
$$y_1 = 3419875491033170827167861896082688,$$
$$x_2 = 1415926535897932384626433832795028,$$
$$y_2 = 3846759606494706724286139623885544,$$

show that the following value of k

$$k = 312521636014772477161767351856699$$

is the correct value satisfying

$$Q(x_2, y_2) \equiv k \cdot P(x_1, y_1) \ (\text{mod } p).$$

11. In ECCp-131, given

$$E \backslash \mathbb{F}_p : y^2 \equiv x^3 + ax + b \ (\text{mod } p),$$
$$\{P(x_1, y_1), Q(x_2, y_2)\} \in E(\mathbb{F}_p),$$
$$p = 1550031797834347859248576414813139942411,$$
$$a = 1399267573763578815877905235971153316710,$$
$$b = 1009296542191532464076260367525816293976,$$
$$x_1 = 1317953763239595888465524145589872695690,$$
$$y_1 = 434829348619031278460656303481105428081,$$
$$x_2 = 1247392211317907151303247721489640699240,$$
$$y_2 = 207534858442090452193999571026315995117,$$

find the correct value of k such that

$$Q(x_2, y_2) \equiv k \cdot P(x_1, y_1) \ (\text{mod } p).$$

4.2 ECDLP-Based Cryptography

Basic Ideas in ECDLP-Based Cryptography

Since ECDLP is also computationally infeasible in polynomial time, it can thus be used to construct unbreakable cryptographic systems:

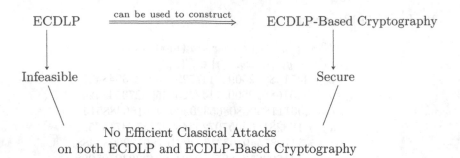

The first two people to ECDLP to construct cryptographic systems, now widely known as elliptic curve cryptography (ECC), are Miller and Koblitz, who independently published the following two seminal papers:

1. Victor Miller [46], "Uses of Elliptic Curves in Cryptography", *Lecture Notes in Computer Science* **218**, Springer, 1986, pp 417–426.

2. Neal Koblitz [31]:1987ECC, "Elliptic Curve Cryptography", *Math. Comput.*, **48** (1987), pp 203–209.

Since then, ECDLP and ECC have been studied extensively, and many practical elliptic curve cryptographic systems and protocols have been developed. Today, ECC is a standard term in the field.

Precomputations of ECC

To implement ECC, we need to do the following precomputations:

1. Embed messages on elliptic curves: Our aim here is to do cryptography with elliptic curve groups in place of $\mathbb{F}_q$. More specifically, we wish to embed plaintext messages as points on an elliptic curve defined over a finite field $\mathbb{F}_q$, with $q = p^r$ and $p \in$ primes. Let our message units m be integers $0 \leqslant m \leqslant M$, and let also κ be a large enough integer for us to be satisfied with an error probability of $2^{-\kappa}$ when we attempt to embed a plaintext message m. In practice, $30 \leqslant \kappa \leqslant 50$. Now let us take $\kappa = 30$ and an elliptic curve $E : y^2 = x^3 + ax + b$ over $\mathbb{F}_q$. Given a message number m, we compute a set of values for x:

$$x = \{m\kappa + j, \ j = 0, 1, 2, \ldots\} = \{30m, \ 30m + 1, \ 30m + 2, \ \cdots\}$$

until we find that $x^3 + ax + b$ is a square modulo p, giving us a point $(x, \sqrt{x^3 + ax + b})$ on E. To convert a point (x, y) on E back to a message number m, we just compute $m = \lfloor x/30 \rfloor$. Since $x^3 + ax + b$ is a square for approximately 50% of all x, there is only about a $2^{-\kappa}$ probability that this method will fail to produce a point on E over $\mathbb{F}_q$. In what follows, we shall give a simple example of how to embed a message number by a point on an elliptic curve. Let E be $y^2 = x^3 + 3x$, $m = 2174$, and $p = 4177$ (in practice, we select $p > 30m$). Then we calculate $x = \{30 \cdot 2174 + j, \ j = 0, 1, 2, \ldots\}$ until $x^3 + 3x$ is a square modulo 4177. We find that when $j = 15$:

$$
\begin{aligned}
x &= 30 \cdot 2174 + 15 \\[4pt]
&= 65235, \\[4pt]
x^3 + 3x &= (30 \cdot 2174 + 15)^3 + 3(30 \cdot 2174 + 15) \\[4pt]
&= 277614407048580 \\[4pt]
&\equiv 1444 \bmod 4177 \\[4pt]
&\equiv 38^2.
\end{aligned}
$$

So we get the message point for $m = 2174$:

$$
(x, \ \sqrt{x^3 + ax + b}) = (65235, 38).
$$

To convert the message point $(65235, 38)$ on E back to its original message number m, we just compute

$$
m = \lfloor 65235/30 \rfloor = \lfloor 2174.5 \rfloor = 2174.
$$

2. Multiply points on elliptic curves over $\mathbb{F}_q$: We have discussed the calculation of $kP \in E$ over $\mathbb{Z}/n\mathbb{Z}$. In elliptic curve public-key cryptography, we are now interested in the calculation of $kP \in E$ over $\mathbb{F}_q$, which can be done in $\mathcal{O}(\log k (\log q)^3)$ bit operations by the *repeated doubling method*. If we happen to know N, the number of points on our elliptic curve E, and if $k > N$, then the coordinates of kP on E can be computed in $\mathcal{O}((\log q)^4)$ bit operations; recall that the number N of points on E satisfies $N \leqslant q + 1 + 2\sqrt{q} = \mathcal{O}(q)$ and can be computed by René Schoof's algorithm in $\mathcal{O}((\log q)^8)$ bit operations.

3. Compute elliptic curve discrete logarithms: Let E be an elliptic curve over $\mathbb{F}_q$ and B a point on E. Then the *discrete logarithm* on E is the problem; given a point $P \in E$, find an integer $x \in \mathbb{Z}$ such that $xB = P$ if such an integer x exists. It is likely that the DLP on elliptic curves over $\mathbb{F}_q$ is more intractable than the DLP in $\mathbb{F}_q$. It is this feature that makes cryptographic systems based on elliptic curves even more secure than that based on the DLP. In the rest of this section, we shall discuss elliptic curve analogues of some important public-key cryptosystems.

In what follows, we shall present some elliptic curve analogues of four widely used public-key cryptosystems, namely, the elliptic curve DHM, the elliptic curve Massey–Omura, the elliptic curve ElGamal, the elliptic curve RSA, and ECDSA.

Elliptic Curve DHM

The Diffie–Hellman–Merkle key-exchange scheme over a finite field $\mathbb{F}_p$ can be easily extended to elliptic curve E over a finite field $\mathbb{F}_p$ (denoted by $E\backslash(\mathbb{F}_p)$); such an elliptic curve analog may be described as follows (see Fig. 4.2).

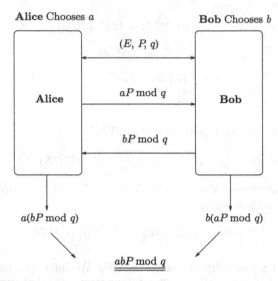

Figure 4.2. Elliptic curve DHM key-exchange scheme

1. Alice and Bob publicly choose a finite field $\mathbb{F}_q$ with $q = p^r$ and $p \in$ primes, an elliptic curve E over $\mathbb{F}_q$, and a random *base* point $P \in E$ such that P generates a large subgroup of E, preferably of the same size as that of E itself. All of this is public information.

2. To agree on a secret key, Alice and Bob choose two secret random integers a and b. Alice computes $aP \in E$ and sends aP to Bob; Bob computes $bP \in E$ and sends bP to Alice. Both aP and bP are, of course, public, but a and b are not.

3. Now both Alice and Bob compute the secret key $abP \in E$, and use it for further secure communications.

4. Cryptanalysis: For the eavesdropper Eve to get abP, she has to find either a from (abP, P) or b from (bP, P).

As everybody knows, there is no known fast way to compute abP if one only knows P, aP, and bP—this is the infeasible ECDLP.

Example 4.4. The following is an elliptic curve analog of the DHM scheme. Let

$$E \backslash \mathbb{F}_{199} : \ y^2 \equiv x^3 + x - 3,$$
$$P = (1, 76) \in E(\mathbb{F}_{199}),$$
$$a = 23,$$
$$b = 86.$$

Then

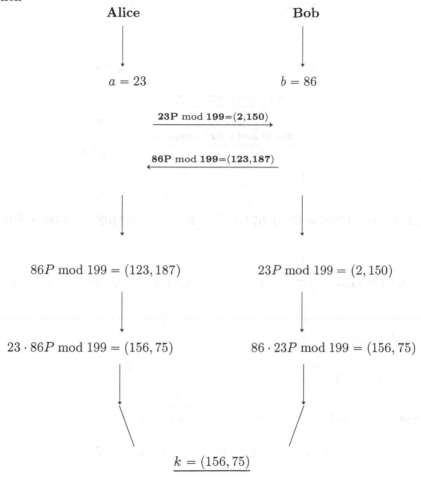

Alice	Bob
$a = 23$	$b = 86$

23P mod 199=(2,150) →

← 86P mod 199=(123,187)

$86P \bmod 199 = (123, 187)$ $23P \bmod 199 = (2, 150)$

$23 \cdot 86P \bmod 199 = (156, 75)$ $86 \cdot 23P \bmod 199 = (156, 75)$

$$k = (156, 75)$$

Clearly, anyone who can find the discrete logarithm a or b such that

$$(2, 150) \equiv a(1, 76) \ (\text{mod } 199), \quad (123, 187) \equiv b(1, 76) \ (\text{mod } 199)$$

can get the key $abP \equiv (156, 75) \ (\text{mod } 199)$.

Example 4.5. We illustrate another example of the elliptic curve analog of the DHM scheme. Let

$$E\backslash\mathbb{F}_{11027} : \ y^2 \equiv x^3 + 4601x + 548,$$
$$P = (9954, 8879) \in E(\mathbb{F}_{11027}),$$
$$a = 1374,$$
$$b = 2493.$$

Then

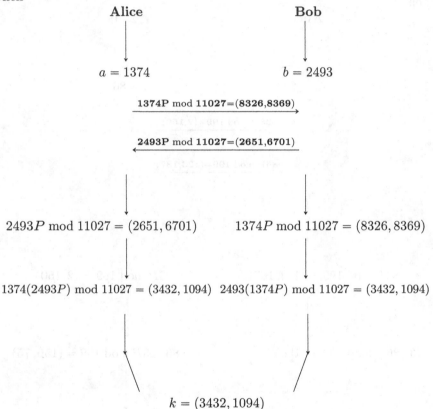

Alice   Bob

$a = 1374$ $b = 2493$

1374P mod 11027=(8326,8369)

2493P mod 11027=(2651,6701)

$2493P \bmod 11027 = (2651, 6701)$ $1374P \bmod 11027 = (8326, 8369)$

$1374(2493P) \bmod 11027 = (3432, 1094)$ $2493(1374P) \bmod 11027 = (3432, 1094)$

$k = (3432, 1094)$

Anyone who can find the discrete logarithm a or b such that

$$(8326, 8369) \equiv a(9954, 8879) \ (\bmod\ 11027)$$

or

$$(2651, 6701) \equiv b(9954, 8879) \ (\bmod\ 11027)$$

can get the key $abP \equiv ((3432, 1094)) \ (\bmod\ 11027)$.

Elliptic Curve Massey–Omura

Recall that the Massey–Omura cryptographic scheme is a three-pass protocol for sending messages, allowing Alice to securely send a message to Bob without the need to exchange or distribute encryption keys. Let E be an elliptic curve over $\mathbb{F}_q$ with q, a prime power and $M = P \in E(\mathbb{F}_q)$, the original message point. Then the elliptic curve analog of the Massey–Omura cryptosystem may be described as follows (see also Fig. 4.3).

1. Alice and Bob publicly choose an elliptic curve E over $\mathbb{F}_q$ with $q = p^r$, a large prime power; as usual, we assume $q = p$ and we suppose also that the number of points on $E \backslash \mathbb{F}_q$ (denoted by N) is publicly known.
2. Alice chooses a secret pair of numbers (e_A, d_A) such that $d_A e_A \equiv 1$ (mod N). Similarly, Bob chooses (e_B, d_B) such that $d_B e_B \equiv 1 \pmod N$.
3. If Alice wants to send a secret message-point $P \in E$ to Bob, then the procedure should be as follows:
   (a) Alice sends $e_A P \bmod q$ to Bob.
   (b) Bob sends $e_B e_A P \bmod q$ to Alice.
   (c) Alice sends $d_A e_B e_A P \bmod q = e_B P$ to Bob.
   (d) Bob computes $d_B e_B P = P$ and hence recovers the original message point.

Note that an eavesdropper would know $e_A P$, $e_B e_A P$, and $e_B P$. So if he could solve the ECDLP on E, he could determine e_B from the first two points and then compute $d_B = e_B^{-1} \bmod q$ and hence get $P = d_B(e_B P)$.

Example 4.6. We follow closely the steps in the above-discussed elliptic curve Massey–Omura cryptography. Let

$$p = 13,$$
$$E \backslash \mathbb{F}_{13} : \ y^2 \equiv x^3 + 4x + 4 \pmod{13}$$
$$|E(\mathbb{F}_{13})| = 15$$
$$M = (12, 8)),$$
$$(e_A, d_A) \equiv (7, 13) \pmod{15},$$
$$(e_B, d_B) \equiv (2, 8) \pmod{15}.$$

Then

$$e_A M \equiv 7(12, 8) \pmod{13},$$
$$\equiv (1, 10) \pmod{13},$$
$$e_A e_B M \equiv e_B(1, 10) \pmod{13},$$

$M \in E(\mathbb{F}_q)$,

$|E(\mathbb{F}_q)| = N$

Alice generates (e_A, d_A) such that $e_A d_A \equiv 1 \pmod{N}$ and sends e_A to Bob.

Bob generates (e_B, d_B) such that $e_B d_B \equiv 1 \pmod{N}$ and sends e_B to Alice.

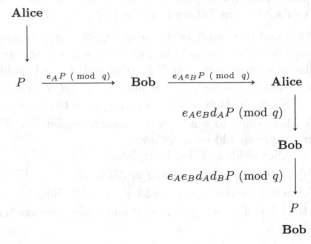

Figure 4.3. The Massey–Omura cryptography

$$\equiv 2(1, 10) \pmod{13},$$

$$\equiv (12, 5) \pmod{13},$$

$$e_A e_B d_A M \equiv d_A(12, 5) \pmod{13},$$

$$\equiv 13(12, 5) \pmod{13},$$

$$\equiv (6, 6) \pmod{13},$$

$$e_A e_B d_A d_B M \equiv d_B(6, 6) \pmod{13},$$

$$\equiv 8(6, 6) \pmod{13},$$

$$\equiv (12, 8) \pmod{13},$$

$$\downarrow$$

$$M.$$

Example 4.7. Let

$$p = 13,$$

$$E\backslash \mathbb{F}_{13}: \ y^2 \equiv x^3 + x \pmod{13}$$

$$|E(\mathbb{F}_{13})| = 20$$
$$M = (11, 9)),$$
$$(e_A, d_A) \equiv (3, 7) \ (\text{mod } 20),$$
$$(e_B, d_B) \equiv (13, 17) \ (\text{mod } 20).$$

Then

$$e_A M \equiv 3(11, 9) \ (\text{mod } 13),$$
$$\equiv (7, 5) \ (\text{mod } 13),$$
$$e_A e_B M \equiv e_B(7, 5) \ (\text{mod } 13),$$
$$\equiv 13(7, 5) \ (\text{mod } 13),$$
$$\equiv (11, 4) \ (\text{mod } 13),$$
$$e_A e_B d_A M \equiv d_A(11, 4) \ (\text{mod } 13),$$
$$\equiv 17(11, 4) \ (\text{mod } 13),$$
$$\equiv (7, 5) \ (\text{mod } 13),$$
$$e_A e_B d_A d_B M \equiv d_B(7, 5) \ (\text{mod } 13),$$
$$\equiv 17(7, 5) \ (\text{mod } 13),$$
$$\equiv (11, 9) \ (\text{mod } 13),$$
$$\downarrow$$
$$M.$$

Elliptic Curve ElGamal

Just the same as many other public-key cryptosystems, the famous ElGamal cryptosystem also has a very straightforward elliptic curve analog, which may be described as follows (see also Fig. 4.4).

1. Suppose Bob wishes to send a secret message to Alice:

$$\text{Bob} \xrightarrow{\text{Secret Message}} \text{Alice}.$$

Alice and Bob publicly choose an elliptic curve E over $\mathbb{F}_q$ with $q = p^r$, a prime power, and a random *base* point $P \in E$. Suppose they also know the number of points on E, i.e., they know $|E(\mathbb{F}_q)| = N$.

2. Alice chooses a random integer a, computes $aP \bmod q$, and sends it to Bob.

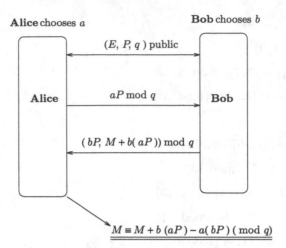

Figure 4.4. Elliptic curve ElGamal cryptography

3. Encryption: Bob chooses at random an integer b and computes $bP \bmod q$. Bob also computes $M + b(aP) \bmod q$. Then Bob sends the secret encrypted message $(bP, M + b(aP)) \bmod q$ to Alice.

4. Decryption: Since Alice has the secret key a, she can compute $a(bP) \bmod q$ and get

$$M \equiv M + a(bP) - b(aP) \pmod{q}, \qquad (4.20)$$

the original plaintext message.

[4] Cryptanalysis: Eve, the eavesdropper, can only get M if she can solve the ECDLP. That is, she can get M if she can find a from $aP \bmod q$ or b from $bP \bmod q$. But as everybody knows, there is no efficient way to compute the elliptic curve discrete logarithms, so the ElGamal cryptosystem system is secure.

Example 4.8. Suppose Bob wishes to send Alice a secret message M by using the elliptic curve ElGamal cryptographic scheme.

[1] Setup:

$$E \backslash \mathbb{F}_{29} : \ y^2 \equiv x^3 - x + 16 \pmod{29}.$$

$$N = |E(\mathbb{F}_{29})| = 31.$$

$$P = (5, 7) \in E(\mathbb{F}_{29}).$$

$$M = (28, 25).$$

[2] Public-key generation: Assume Bob sends the secret message M to Alice, so Alice:

> Chooses a random secret integer $a = 23$
> Computes $aP = 23P = (21, 18) \pmod{29}$
> Sends $aP = (21, 18) \pmod{29}$ to Bob

[3] Encryption: Bob

> Chooses a random secret integer $b = 25$
> Computes $bP = 25P = (13, 24) \pmod{29}$,
> $\qquad b(aP) = 17(23P) = 17(21, 18) = (1, 25) \pmod{29}$,
> $\qquad M + b(aP) = (28, 25) + (1, 25) = (0, 4) \pmod{29}$
> Sends $(bP = (1, 25),\ M + b(aP) = (0, 4))$ to Alice

[4] Decryption: Alice computes

$$a(bP) = 23(25P) = 23(13, 24) = (1, 25),$$
$$\begin{aligned} M &= M + b(aP) - a(bP) \\ &= (0, 4) - (1, 25) \\ &= (0, 4) + (1, -25) \\ &= (28, 25). \end{aligned}$$

So, Alice recovers the original secret message $M = (28, 25)$.

Example 4.9. Now we give one more example on elliptic curve ElGamal cryptosystem.

[1] Setup:

$$E\backslash \mathbb{F}_{523}:\ y^2 \equiv x^3 + 22x + 153 \pmod{523}.$$

$$P = (167, 118) \in E(\mathbb{F}_{523}).$$

$$M = (220, 287) \text{ is the plaintext.}$$

[2] Public-key generation: Assume Bob sends the secret message M to Alice, so Alice:

> Chooses a random secret integer $a = 97$,
> Computes $aP = 97(167, 118) = (167, 405) \pmod{523}$,
> Sends $aP = (167, 405) \pmod{523}$ to Bob.

[3] Encryption: Bob

> Chooses a random secret integer $b = 263$,
> Computes $bP = 263(167, 118) = (5, 503) \pmod{523}$,
> $\qquad b(aP) = 263(167, 405) = (5, 20) \pmod{523}$,
> $\qquad M + b(aP) = (220, 287) + (5, 20)$
> $\qquad\qquad\qquad = (36, 158) \pmod{523}$,
> Sends $(bP = (5, 503),\ M + b(aP) = (36, 158))$ to Alice.

[4] Decryption: Alice computes

$$a(bP) = 97(5, 503) = (5, 20),$$
$$M = M + b(aP) - a(bP)$$
$$= (36, 158) - (5, 20)$$
$$= (36, 158) + (5, 503)$$
$$= (220, 287).$$

So, Alice recovers the original secret message $M = (220, 287)$.

Elliptic Curve RSA

The most widely used RSA cryptosystem cannot only be used to both en-cryption and digital signatures but also has a natural analog of ECC, and in fact, several elliptic curve RSA cryptosystems have been developed. In what follows, we describe one of the analogues:

1. $N = pq$ is a public-key which is the product of the two large secret primes p and q.

2. Choose two random integers a and b such that $E : y^2 = x^3 + ax + b$ defines an elliptic curve both mod p and mod q.

3. To encrypt a message-point P, just perform eP mod N, where e is the public (encryption) key. To decrypt, one needs to know the number of points on E modulo both p and q.

The above are some elliptic curve analogues of certain public-key cryp-tosystems. It should be noted that almost every public-key cryptosystem has an elliptic curve analogue; it is of course possible to develop new elliptic curve cryptosystems which do not rely on the existing cryptosystems.

It should be also noted that the digital signature schemes can also be ana-logued by elliptic curves over $\mathbb{F}_q$ or over $\mathbb{Z}/n\mathbb{Z}$ with $n = pq$ and $p, q \in$ primes in exactly the same way as that for public-key cryptography; several elliptic curve analogues of digital signature schemes have already been proposed, say, e.g., [45].

Menezes–Vanstone ECC

A serious problem with all above-mentioned elliptic curve cryptosystems is that the plaintext message units m lie on the elliptic curve E and there is no convenient method known of deterministically generating such points on E. Fortunately, Menezes and Vanstone had discovered a more efficient variation [40]; in this variation, which we shall describe below, the elliptic curve is used for "masking", and the plaintext and ciphertext pairs are allowed to be in $\mathbb{F}_p^* \times \mathbb{F}_p^*$ rather than on the elliptic curve.

1. Key generation: Alice and Bob publicly choose an elliptic curve E over $\mathbb{F}_p$ with $p > 3$ is prime and a random *base* point $P \in E(\mathbb{F}_p)$ such that P generates a large subgroup H of $E(\mathbb{F}_p)$, preferably of the same size as that of $E(\mathbb{F}_p)$ itself. Assume that randomly chosen $k \in \mathbb{Z}_{|H|}$ and $a \in \mathbb{N}$ are secret.

2. Encryption: Suppose now Alice wants to send message

$$m = (m_1, m_2) \in (\mathbb{Z}/p\mathbb{Z})^* \times (\mathbb{Z}/p\mathbb{Z})^* \tag{4.21}$$

   to Bob, then she does the following:
   (a) $\beta = aP$, where P and β are public.

   (b) $(y_1, y_2) = k\beta$.

   (c) $c_0 = kP$.

   (d) $c_j \equiv y_j m_j \pmod{p}$ for $j = 1, 2$.

   (e) Alice sends the encrypted message c of m to Bob:

$$c = (c_0, c_1, c_2). \tag{4.22}$$

3. Decryption: Upon receiving Alice's encrypted message c, Bob calculates the following to recover m:
   (a) $ac_0 = (y_1, y_2)$.

   (b) $m = \left(c_1 y_1^{-1} \pmod{p}, \ c_2 y_2^{-1} \pmod{p}\right)$.

Example 4.10. The following is a nice example of Menezes–Vanstone cryptosystem [47].

[1] Key generation: Let E be the elliptic curve given by $y^2 = x^3 + 4x + 4$ over $\mathbb{F}_{13}$ and $P = (1, 3)$ be a point on E. Choose $E(\mathbb{F}_{13}) = H$ which is cyclic of order 15, generated by P. Let also the private-keys $k = 5$ and $a = 2$ and the plaintext $m = (12, 7) = (m_1, m2)$.

[2] Encryption: Alice computes

$$\beta = aP = 2(1, 3) = (12, 8),$$
$$(y_1, y_2) = k\beta = 5(12, 8) = (10, 11),$$
$$c_0 = kP = 5(1, 3) = (10, 2),$$
$$c_1 \equiv y_1 m_1 \equiv 10 \cdot 2 \equiv 3 \pmod{13},$$
$$c_2 \equiv y_2 m_2 \equiv 11 \cdot 7 \equiv 12 \pmod{13}.$$

Then Alice sends

$$c = (c_0, c_1, c_2) = ((10, 2), 3, 12)$$

to Bob.

[3] Decryption: Upon receiving Alice's message, Bob computes

$$ac_0 = 2(10, 2) = (10, 11) = (y_1, y_2),$$
$$m_1 \equiv c_1 y_1^{-1} \equiv 12 \pmod{13},$$
$$m_2 \equiv c_2 y_2^{-1} \equiv 7 \pmod{13}.$$

Thus, Bob recovers the message $m = (12, 7)$.

Elliptic Curve DSA

We have already noted that almost every public-key cryptosystem has an elliptic curve analogue. It should also be noted that digital signature schemes can also be represented by elliptic curves over $\mathbb{F}_q$ with q, a prime power, or over $\mathbb{Z}/n\mathbb{Z}$ with $n = pq$ and $p, q \in$ primes. In exactly the same way as that for public-key cryptography, several elliptic curve analogues of digital signature schemes have already been proposed (see, for example, Meyer and Müller [45]). In what follows we shall describe an elliptic curve analogue of the DSA/DSS, called ECDSA [26].

Algorithm 4.4 (ECDSA). Let E be an elliptic curve over $\mathbb{F}_p$ with p prime, and let P be a point of prime order q (note that the q here is just a prime number, not a prime power) in $E(\mathbb{F}_p)$. Suppose Alice wishes to send a signed message to Bob.

[1] [ECDSA key generation] Alice does the following:

   [1-1] select a random integer $x \in [1, \ q - 1]$,

   [1-2] compute $Q = xP$,

   [1-3] make Q public, but keep x as a secret.
   Now Alice has generated the public key Q and the private-key x.

[2] [ECDSA signature generation] To sign a message m, Alice does the following:

   [2-1] select a random integer $k \in [1, \ q - 1]$,

   [2-2] compute $kP = (x_1, y_1)$, and $r \equiv x_1 \pmod{q}$. If $r = 0$, go to step [2-1].

   [2-3] compute $k^{-1} \bmod q$.

   [2-4] compute $s \equiv k^{-1}(H(m) + xr) \pmod{q}$, where $H(m)$ is the hash value of the message. If $s = 0$, go to step [2-1].
   The signature for the message m is the pair of integers (r, s).

[3] [ECDSA signature verification] To verify Alice's signature (r, s) of the message m, Bob should do the following:

[3-1] obtain an authenticated copy of Alice's public key Q;

[3-2] verify that (r, s) are integers in the interval $[1, q - 1]$, computes $kP = (x_1, y_1)$, and $r \equiv x_1 \pmod{q}$.

[3-3] compute $w \equiv s^{-1} \pmod{q}$ and $H(m)$.

[3-4] compute $u_1 \equiv H(m)w \pmod{q}$ and $u_2 \equiv rw \pmod{q}$.

[3-5] compute $u_1 P + u_2 Q = (x_0, y_0)$ and $v \equiv x_0 \pmod{q}$.

[3-6] accept the signature if and only if $v = r$.

As a conclusion to ECC, we provide two remarks about the comparison of ECC and other types of cryptography, particularly the famous and widely used RSA cryptography.

Remark 4.1. ECC provides a high level of security using smaller keys than that used in RSA. A comparison between the key sizes for an equivalent level of security for RSA and ECC is given in the following Table 4.2.

Table 4.2. Key size comparison between RSA and ECC

Security level	RSA	ECC
Low	512 bits	112 bits
Medium	1024 bits	161 bits
High	3027 bits	256 bits
Very high	15360 bits	512 bits

Remark 4.2. Just the same that there are weak keys for RSA, there are also weak keys for ECC, say, for example, as an acceptable elliptic curve for cryptography, it must satisfy the following conditions:

1. If N is the number of integer coordinates, it must be divisible by a large prime r such that $N = kr$ for some integer k.

2. If the curve has order p modulo p, then r must not be divisible by $p^i - 1$ for a small set of i, say, $0 \leqslant i \leqslant 20$.

3. Let N be the number of integer coordinates and $p = E(\mathbb{F}_p)$, and then N must not equal to p. The curve that satisfies the condition $p = N$ is called the anomalous curve.

Exercises and Problems for Sect. 4.2

1. Describe the advantages of ECC over integer factoring-based and discrete logarithm-based cryptography.

2. Give the complexity measures for the fastest known general algorithms for:
   (a) The IFP
   (b) The DLP
   (c) The ECDLP

3. Give the complexity measures for
   (a) The IFP based cryptosystems
   (b) The DLP based cryptosystems
   (c) The ECDLP based cryptosystems

4. The exponential cipher, invented by Pohlig and Hellman in 1978 and based on the mod p arithmetic, is a secret-key cryptosystem, but it is very close to the RSA public-key cryptosystem based on mod n arithmetic, where $n = pq$ with p, q prime numbers. In essence, the Pohlig–Hellman cryptosystem works as follows:
   (a) Choose a large prime number p and the encryption key k such that $0 < k < p$ and $\gcd(k, p-1) = 1$.
   (b) Compute the decryption key k' such that $k \cdot k' \equiv 1 \pmod{p-1}$.
   (c) Encryption: $C \equiv M^k \pmod{p}$.
   (d) Decryption: $M \equiv C^{k'} \pmod{p}$.
   Clearly, if you change the modulo p to modulo $n = pq$, then the Pohlig–Hellman cryptosystem is just the RSA cryptosystem.
   (a) Design an elliptic curve analog of the Pohlig–Hellman cryptosystem.

   (b) Explain why the original Pohlig–Hellman cryptosystem is easy to break, whereas the elliptic curve Pohlig–Hellman cryptosystem is hard to break.

5. Koyama et al. [36] proposed three trap-door one-way functions; one of the functions claimed to be applicable to zero-knowledge identification protocols. Give an implementation of the elliptic curve trap-door one-way function to the zero-knowledge identification protocol.

6. Suppose that Alice and Bob want to establish a secret key for future encryption in EC DHM key exchange. Both Alice and Bob perform as follows:

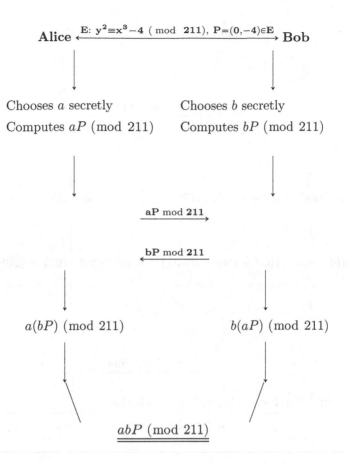

Find the actual values for

(a) aP mod 211

(b) bP mod 211

(c) abP mod 211

(d) baP mod 211

Verify if $abP \equiv baP \pmod{211}$.

7. Let the elliptic curve analog of a DHM scheme be as follows.

$$E\backslash \mathbb{F}_{11027} : \ y^2 \equiv x^3 + 4601x + 548,$$

$$P = (2651, 6701) \in E(\mathbb{F}_{11027}),$$

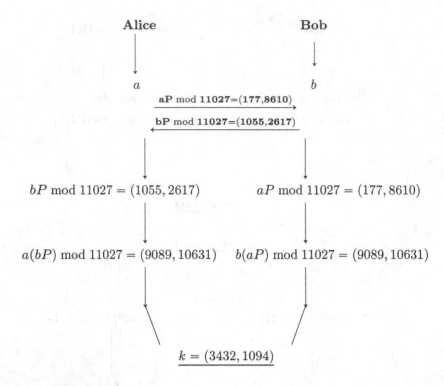

(a) Find the discrete logarithm a such that
$$aP \bmod 11027 = (177, 8610).$$

(b) Find the discrete logarithm b such that
$$bP \bmod 11027 = (1055, 2617).$$

8. Consider the elliptic curve E
$$E : \ y^2 = x^3 + x - 3$$
over the field $\mathbb{F}_{199}$. Let $M = (1, 76) \in E(\mathbb{F}_{199})$ and $(e_A, e_B) = (23, 71)$.
(a) Find the number of points, N, in $E(\mathbb{F}_{199})$.
(b) Find
$$e_A P \bmod q,$$
$$e_A e_B M \bmod q.$$

    (c) Find
$$e_A e_B d_A M \bmod q,$$
$$e_A e_B d_A d_B M \bmod q.$$
    (d) Check if $e_A e_B d_A d_B M \bmod q = P$?

9. Consider the elliptic curve E

$$E : \; y^2 = x^3 + 1441x + 611$$

over the field $\mathbb{F}_{2591}$. Let $P = (1619, 2103) \in E(\mathbb{F}_{2591})$, $(e_A, e_B) = (107, 257)$.

    (a) Find the number of points, N, in $E(\mathbb{F}_{2591})$.

    (b) Find
$$e_A P \bmod q,$$
$$e_A(e_B M) \bmod q.$$

    (c) Find
$$d_A(e_A e_B)M \bmod q,$$
$$d_B(d_A e_A e_B M) \bmod q.$$
    (d) Check if $e_A e_B d_A d_B P \bmod q = M$?

10. Let p be a 200-digit prime number as follows:

    $p =$ 1000000000000000000000000000000000000000000000000000_
        0000000000000000000000000000000000000000000000000000_
        0000000000000000000000000000000000000000000000000000_
        0000000000000000000000000000000000000000000000000153

Let the elliptic curve over $\mathbb{F}_p$ be as follows:

$$E\backslash\mathbb{F}_p : \; y^2 \equiv x^3 + 105x + 78153 \pmod{p},$$

with a point order:

    $N =$ 1000000000000000000000000000000000000000000000000000_
        0000000000000000000000000000000000000000000000000000_
        06789750288004224118080314365460277641928049641888_
        399915913929600322106305617600290508586136896316753

    (a) Let $e_A = 179$, compute $d_A \equiv \frac{1}{e_A} \bmod N$.

    (b) Let $e_B = 983$, compute $d_B \equiv \frac{1}{e_B} \bmod N$.

11. Let p be a prime number

    $p =$ 123456789012345678901234567890654833374525085966673_
        7125236501

Let the elliptic curve over $\mathbb{F}_p$ be as follows:
$$y^2 \equiv x^3 +$$
1125079135286236108376138855036822306988688835725996813843335x
$-$1125079135286236108376138855036822306988688835725996813843335
(mod p)
with order $|E(\mathbb{F}_p)| = N$ as follows:

  1234567890123456789012345678901234567890123456789012345678901234568197.

Let
(7642989232975292895356351754903278029804860223284406315749,
10018174132244810544452087161446405316940052977694565571441)
be the plaintext point M. Suppose Alice wishes to send M to Bob.
Let

$e_A = 3,$
$d_A = 8230452600082304526008230452600823045260082304526008230452600823045465,$
$e_B = 7,$
$d_B = 17636684144620811271604938270017636684144620811271604938314,$

all modulo N. Compute
(a) $e_A M \bmod p$
(b) $e_B(e_A M) \bmod p$
(c) $d_A(e_B e_A M) \bmod p$
(d) $d_B(d_A e_B e_A M) \bmod p$
Check if $d_B(d_A e_B e_A M) \bmod p = M$.

12. Suppose that Alice wants to send Bob a secret massage $M = (10, 9)$ using elliptic curve ElGamal cryptography. Both Alice and Bob perform as follows:

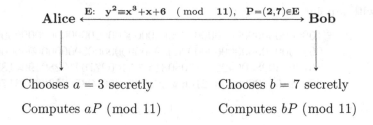

Alice ←——— E: $y^2 \equiv x^3 + x + 6$ (mod 11), P=(2,7)∈E ———→ Bob

Chooses $a = 3$ secretly   Chooses $b = 7$ secretly

Computes aP (mod 11)   Computes bP (mod 11)

bP mod 11

{aP, M+a(bP)} mod 11

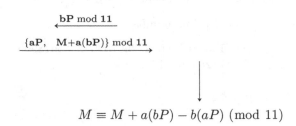

$$M \equiv M + a(bP) - b(aP) \pmod{11}$$

Compute the actual values for
(a) $aP \bmod 11$
(b) $bP \bmod 11$
(c) $b(aP) \bmod 11$
(d) $a(bP) \bmod 11$
(e) $M + a(bP) \bmod 11$
(f) $M + a(bP) - b(aP) \pmod{11}$
Check if $M + a(bP) - b(aP) \pmod{11} = (10, 9)$.

13. Suppose that Alice wants to send Bob a secret massage $M = (562, 201)$ in elliptic curve ElGamal cryptography. Both Alice and Bob perform the following:

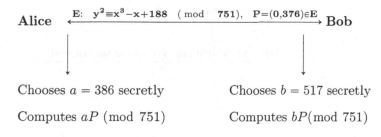

$$M \equiv M + a(bP) - b(aP) \pmod{751}$$

Compute the actual values for
(a) $aP \bmod 751$
(b) $bP \bmod 751$
(c) $a(bP) \bmod 751$
(d) $b(aP) \bmod 751$
(e) $M + a(bP) \bmod 751$
(f) $M + a(bP) - b(aP) \bmod 751$
Check if $M + a(bP) - b(aP) \bmod 751 = (562, 201)$.

14. Suppose that Alice wants to send Bob a secret massage $M = (316, 521)$ in elliptic curve ElGamal cryptography. Both Alice and Bob performs the following:

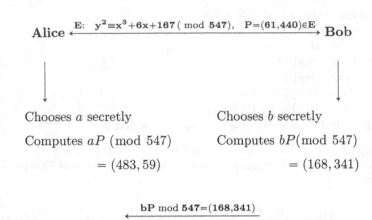

Alice $\xleftrightarrow{\quad \text{E:} \quad y^2 \equiv x^3+6x+167\,(\bmod\ 547), \quad P=(61,440)\in E \quad}$ Bob

Chooses a secretly   Chooses b secretly

Computes aP (mod 547)   Computes bP(mod 547)

$= (483, 59)$ $= (168, 341)$

$\xleftarrow{\qquad \text{bP mod 547}=(168,341) \qquad}$

$\xrightarrow{\qquad \{\text{aP}, \quad \text{M}+\text{a(bP)}\}\ \bmod\ 547=\{(483,59),(49,178)\} \qquad}$

$$M \equiv M + a(bP) - b(aP) \ (\bmod\ 547)$$

$$\equiv (49, 178) + (143, -443) \ (\bmod\ 547)$$

$$\equiv (316, 521) \ (\bmod\ 547)$$

Find

(a) a such that aP mod 547 $= (483, 59)$.

(b) b such that bP mod 547 $= (168, 341)$.

(c) $a(bP)$ mod 547.

(d) $b(aP)$ mod 547.

Check if $a(bP) \equiv b(aP) \ (\bmod\ 547)$.

15. Let $E \backslash \mathbb{F}_{2^m}$ be the elliptic curve E over $\mathbb{F}_{2^m}$ with $m > 1$, where E is defined to be
$$y^2 + xy = x^3 + ax^2 + b.$$

(a) Let $P, Q \in E$ with $P \neq \pm Q$ are two points on E. Find the addition formula for computing $P + Q$.

(b) Let $P \in E$ with $P \neq -P$. Find the addition formula for computing $2P$.

(c) Let $E\backslash\mathbb{F}_{2^m}$ be as follows:

$$E\backslash\mathbb{F}_{2^4} : \ y^2 \equiv x^3 + \alpha^4 x^2 + 1 \ (\mathrm{mod}\ 2^4).$$

Find all the points, $E(\mathbb{F}_{2^4})$, including the point at infinity, on the E.

(d) Let $P = (\alpha^6, \alpha^8)$ and $Q = (\alpha^3, \alpha^{13})$ be in $E\backslash\mathbb{F}_{2^4}$ defined above; find $P + Q$ and $2P$.

16. Show that breaking ECC or any ECDLP-based cryptography is generally equivalent to solving the ECDLP problem.

4.3 Quantum Attack on ECDLP-Based Cryptography

Basic Idea of Quantum Attacks on ECDLP/ECDLP-Based Cryptography

Shor's quantum algorithms for discrete logarithms can be used to solve the elliptic curve discrete logarithms in $\mathcal{BQP}$.

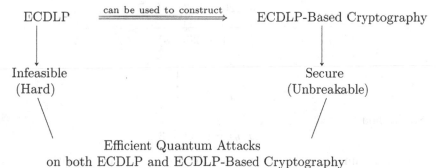

Surprisingly,

Quantum Period Finding Algorithm

$\downarrow$

Quantum ECDLP Algorithm

$\downarrow$

Quantum Attacks on ECDLP-Based Cryptography

As we mentioned earlier, the DLP problem is just the inverse problem—finding the multiplicative inverse in $\mathbb{Z}_p^*$. Remarkably enough, the ECDLP problem is also an inverse problem—finding the additive inverse in $E(\mathbb{F}_p)$. More importantly, the method for solving such an inverse problem is still the Euclid's algorithm, but an elliptic curve version of the old Euclid's and efficient algorithm. Let us first review how Euclid's algorithm can be used to solve (x, y) in the following congruence:

$$ax - by = 1. \tag{4.23}$$

To be more specific, we show how to use the Euclid's algorithm to find x, y in

$$7x - 26y = 1, \tag{4.24}$$

which is equivalent to find x in

$$\frac{1}{7} \equiv x \pmod{26}. \tag{4.25}$$

$$
\begin{aligned}
26 &= 7 \cdot 3 + 5 \quad \rightarrow 5 = 26 - 7 \cdot 3 \\
7 &= 5 \cdot 1 + 2 \quad \rightarrow 2 = 7 - 5 \cdot 1 \\
5 &= 2 \cdot 2 + 1 \quad \rightarrow 1 = 5 - 2 \cdot 2 \\
&= 5 - 2(7 - 5 \cdot 1) \\
&= 3 \cdot 5 - 2 \cdot 7 \\
&= 3 \cdot (26 - 7 \cdot 3) - 2 \cdot 7 \\
&= 3 \cdot 26 - 7 \cdot 11 \\
&= 7(-11) - 26(-3) \\
& \qquad \quad \downarrow \qquad \quad \downarrow \\
& \qquad \quad x \qquad \quad y
\end{aligned}
$$

So, we find

$$(x, y) = (-11, -3).$$

The quantum algorithms, say, e.g., the Proos–Zalka's algorithm [50] and Eicher–Opoku's algorithm [14] for ECDLP, aim at finding (a, b) in

$$aP + bQ = 1. \tag{4.26}$$

Recall that the ECDLP problem asks to find r such that

$$Q = rP,$$

where P is a point of order m on an elliptic curve over a finite field $\mathbb{F}_p$, $Q \in G$ and $G = \langle P \rangle$. A way to find r is to find distinct pairs (a', b') and (a'', b'') of integers modulo r such that

$$a'P + b' = a''P + b''Q.$$

Then

$$(a' - a'')P = (b'' - b')Q,$$

that is,

$$Q = \frac{a' - a''}{b'' - b'}P,$$

or, alternatively,

$$r \equiv \frac{a' - a''}{b'' - b'} \pmod{m}.$$

The computation to find say, e.g., aP can be done efficiently as follows. Let $e_{\beta-1}e_{\beta-2}\cdots e_1e_0$ be the binary representation of a. Then for i starting from $e_{\beta-1}$ down to e_0 ($e_{\beta-1}$ is always 1 and used for initialization), check whether or not $e_i = 1$. If $e_i = 1$, then perform a doubling and an addition group operation; otherwise, just perform a doubling operation. For example, to compute $89P$, since $89 = 1011001$, we have:

e_6	1	P	initialization
e_5	0	$2P$	doubling
e_4	1	$2(2P) + P$	doubling and addition
e_3	1	$2(2(2P) + P) + P$	doubling and addition
e_2	0	$2(2(2(2P) + P) + P)$	doubling
e_1	0	$2(2(2(2(2P) + P) + P))$	doubling
e_0	1	$2(2(2(2(2(2P) + P) + P))) + P$	doubling and addition

$$\|$$
$$89P$$

The following algorithm implements this idea of *repeated doubling and addition* for computing kP.

Algorithm 4.5 (Fast group operations kP on elliptic curves). This algorithm computes aP, where a is a large integer and P is assumed to be a point on an elliptic curve $E: y^2 = x^3 + ax + b$.

[1] Write a in the binary expansion form $a = e_{\beta-1}e_{\beta-2}\cdots e_1e_0$, where each e_i is either 1 or 0. (Assume a has β bits.)

[2] Set $c \leftarrow 0$.

[3] Compute aP:

> for i from $\beta - 1$ down to 0 do
> $\quad c \leftarrow 2c$ (doubling);
> $\quad$ if $e_i = 1$ then $c \leftarrow c + P$; (addition)

[4] Print c; (now $c = aP$).

Note that Algorithm 4.5 does not actually calculate the coordinates (x, y) of kP on an elliptic curve

$$E\backslash\mathbb{F}_p : \ y^2 \equiv x^3 + ax + b \pmod{p}.$$

To make Algorithm 4.5 a practically useful algorithm for point additions on an elliptic curve E, we must incorporate the actual coordinate addition $P_3(x_3, y_3) = P_1(x_1, y_1) + P_2(x_2, y_2)$ on E into the algorithm. To do this, we use the following formulas to compute x_3 and y_3 for P_3:

$$(x_3, y_3) = (\lambda^2 - x_1 - x_2, \ \lambda(x_1 - x_3) - y_1),$$

where

$$\lambda = \begin{cases} \dfrac{3x_1^2 + a}{2y_1} & \text{if } P_1 = P_2 \\[2mm] \dfrac{y_2 - y_1}{x_2 - x_1} & \text{otherwise.} \end{cases}$$

For curves of the form

$$E\backslash\mathbb{F}_{2^m} : \ y^2 + xy \equiv x^3 + ax + b \pmod{2^m},$$

if $P_1 \neq P_2$, then

$$(x_3, y_3) = (\lambda^2 + \lambda + x_1 + x_2 + a, \ \lambda(x_1 + x_3) + x_3 + y_1),$$

where

$$\lambda = \frac{y_1 + y_2}{x_1 + x_2}.$$

If $P_1 = P_2$, then

$$(x_3, y_3) = (\lambda^2 + \lambda + a, \ x_1^2 + \lambda x_3 + x_3),$$

where

$$\lambda = \frac{x_1 + y_1}{x_1}.$$

Also for curves of the form

$$E\backslash\mathbb{F}_{2^m} : \ y^2 + cy \equiv x^3 + ax + b \pmod{2^m},$$

if $P_1 \neq P_2$, then

$$(x_3, y_3) = (\lambda^2 + x_1 + x_2, \ \lambda(x_1 + x_3) + y_1 + c),$$

where

$$\lambda = \frac{y_1 + y_2}{x_1 + x_2}.$$

If $P_1 = P_2$, then

$$(x_3, y_3) = (\lambda^2, \ \lambda(x_1 + x_3) + y_1 + c,$$

where

$$\lambda = \frac{x_1^2 + a}{c}.$$

In what follows, we shall mainly introduce three types of the quantum attacks on ECDLP/ECC:

1. Proos–Zalka's quantum attack on ECDLP

2. Eicher–Opoku's quantum attack on elliptic curve Massey–Omura cryptosystem

3. CMMP quantum attack on ECC

Eicher–Opoku's Quantum Attack on ECDLP

It is quite straightforward to use Shor's quantum algorithm for DLP [53], discussed in the previous chapter, to solve ECDLP in $\mathcal{BQP}$. The following is a modified version of Shor's algorithm to solve the ECDLP problem over $\mathbb{F}_p$ with p prime power (we assume that N is the order of the point P in $E(\mathbb{F}_p)$), based on Eicher–Opoku [14].

Algorithm 4.6 (Quantum Algorithm for ECDLP). The quantum algorithm tries to find

$$r \equiv \log_P Q \ (\text{mod } p) \tag{4.27}$$

such that

$$Q \equiv rP \ (\text{mod } p), \tag{4.28}$$

where $P, Q \in E(\mathbb{F}_p)$ and N is the order of the point P in $E(\mathbb{F}_p)$.

[1] Initializing three required quantum registers as follows:

$$|\Psi_1\rangle = |\mathcal{O}, \ \mathcal{O}, \ \mathcal{O}\rangle, \tag{4.29}$$

where $\mathcal{O}$ denotes the point at infinity, as defined in the elliptic curve group $E(\mathbb{F}_p)$.

[2] Chose q with $p \leqslant q \leqslant 2p$.

[3] Put in the first two registers of the quantum computer the uniform super-position of all $|a\rangle$ and $|b\rangle$ (mod N), and compute $aP + bQ$ (mod N) in the third register. This leaves the quantum computer in the state $|\Psi_2\rangle$:

$$|\Psi_1\rangle = \frac{1}{N} \sum_{a=0}^{N-1} \sum_{b=0}^{N-1} |a,\ b,\ aP + bQ \text{ (mod } N)\rangle \qquad (4.30)$$

Note that $aP + bQ$ (mod p) can be done efficiently by classical *doubling-addition* method [69].

[4] Use the Fourier transform A_q to map $|a\rangle \to |c\rangle$ and $|b\rangle \to |d\rangle$ with probability amplitude

$$\frac{1}{q} \exp\left(\frac{2\pi i}{q}(ac + bd)\right).$$

Thus, the state $|a, b\rangle$ will be changed to the state:

$$\frac{1}{q} \sum_{c=0}^{q-1} \sum_{d=0}^{q-1} \exp\left(\frac{2\pi i}{q}(ac + bd)\right) |c,\ d\rangle. \qquad (4.31)$$

This leaves the machine in the state $|\Psi_3\rangle$:

$$|\Psi_3\rangle = \frac{1}{Nq} \sum_{a,b=0}^{N-1} \sum_{c,d=0}^{q-1} \exp\left(\frac{2\pi i}{q}(ac + bd)\right) |c,\ d,\ aP + bQ \text{ (mod } N)\rangle. \qquad (4.32)$$

[5] Observe the state of the quantum computer and extract the required information. The probability of observing a state $|c,\ d,\ kP \text{ (mod } N)\rangle$ is

$$\text{Prob}(c, d, kP) = \left| \frac{1}{Nq} \sum_{\substack{a,b \\ a-rb \,\equiv\, k \,(\text{mod } N-1)}} \exp\left(\frac{2\pi i}{q}(ac + bd)\right) \right|^2 \qquad (4.33)$$

where the sum is over all (a, b) such that

$$aP + bQ \equiv kP \text{ (mod } N - 1). \qquad (4.34)$$

[6] Just the same as the quantum algorithm for the DLP problem, use the relation

$$a = rb + k - (p - 1) \left\lfloor \frac{br + k}{p - 1} \right\rfloor \qquad (4.35)$$

to substitute in (4.33) to get the amplitude on $|c, d, kP \text{ (mod } N)\rangle$:

$$\frac{1}{Nq} \sum_{b=0}^{N-1} \exp\left(\frac{2\pi i}{q}\left(brc + kc + bd - c(N-1)\left\lfloor \frac{br + k}{p - 1} \right\rfloor\right)\right). \qquad (4.36)$$

This leaves finally the machine in the state $|\Psi_3\rangle$:

$$\frac{1}{Nq}\sum_{b=0}^{N-1}\exp\left(\frac{2\pi i}{q}\left(brc + kc + bd - c(p-1)\left\lfloor\frac{br+k}{N-1}\right\rfloor\right)\right)$$
$$|c, d, kP \pmod{p}\rangle. \qquad (4.37)$$

The probability of observing the above state $|c, d, kP \pmod{N}\rangle$ is thus:

$$\left|\frac{1}{Nq}\sum_{b=0}^{N-1}\exp\left(\frac{2\pi i}{q}\left(brc + kc + bd - c(p-1)\left\lfloor\frac{br+k}{N}\right\rfloor\right)\right)\right|^2. \qquad (4.38)$$

Since $\exp(2\pi ikc/q)$ does not change the probability, (4.36) can be rewritten algebraically as follows:

$$\left|\frac{1}{Nq}\sum_{b=0}^{N-1}\exp\left(\frac{2\pi i}{q}bT\right)\exp\left(\frac{2\pi i}{q}V\right)\right|^2, \qquad (4.39)$$

where

$$T = rc + d - \frac{r}{p-1}\{cN)\}_q, \qquad (4.40)$$

$$V = \left(\frac{br}{N} - \left\lfloor\frac{br+k}{N}\right\rfloor\right)\{cN)\}_q. \qquad (4.41)$$

The notation $\{\alpha\}_q$ here denotes $\alpha \bmod q$ with $-q/2 < \{\alpha\}_q < q/2$.

[7] Finally, deduce r from (c, d). Let j be the closest integer to T/q and $b \in [0, p-2]$, then

$$|\{T\}_q| = |rc + d - \frac{r}{p-1}\{cN\}_q - jq| \leqslant \frac{1}{2}. \qquad (4.42)$$

Further, if

$$|\{cN\}_q| \leqslant \frac{q}{12}, \qquad (4.43)$$

then

$$|V| \leqslant \frac{q}{12}. \qquad (4.44)$$

Therefore, given (c, d), r can be easily calculated with a high probability.

Remark 4.3. Eicher and Opoku also showed in [14] an example of using the algorithm to break a particular elliptic curve Massey–Omura cryptographic system. More specifically, assume that

$$E\backslash\mathbb{F}_{2^5} : y^2 + y \equiv x^3 \pmod{33},$$

$$\mathbb{F}_{2^5} = \{0, 1, \omega, \omega^2, \omega^3, \ldots, \omega^{30}\}$$

$$N = |\mathbb{F}_{2^5}| = 33,$$

$$P_m = \{\omega^{15}, \omega^{10}\},$$

$$e_A P_m = \{\omega^9, \omega^{14}\},$$

$$e_A e_B P_m = \{\omega^{29}, \omega^{16}\},$$

$$e_A e_B d_A P_m = e_B P_m = \{\omega^{18}, \omega^{20}\}.$$

They then give a demonstration of how to use the quantum algorithm to find e_A, since once e_A can be found, $d_B \equiv e_A^{-1} \pmod{33}$ can be found; therefore, $P_m = d_A e_A P_m$, the original message point, can be found.

Proos–Zalka's Quantum Attack on ECDLP Over $\mathbb{F}_p$

Proos and Zalka [50] proposed a quantum algorithm for solving the ECDLP problem over the finite field $\mathbb{F}_p$ with p prime (not equally important to that over the finite field $\mathbb{F}_{2^m}$ or other finite fields). Their experience showed that a smaller quantum computer can break an ECDLP-based cryptographic system with the same level of security of an IFP-based cryptographic system that would need a large computer. More specifically, A 160-bit ECC key could be broken on a quantum computer with about 1000 qubits, whereas factoring the security equivalent 1024-bit RSA modulus would need about 2000 qubits. This means that in classical computation, ECC provides a high level of security using smaller keys than that used in RSA, say, for example, for the same level of security, if an RSA key is about 15360 bits, an ECC key would only need 512 bits. However, in quantum computation, the situation is completely opposite; ECDLP-based cryptography is easy to break than IFP-based cryptography.

In Proos–Zalka's modification of Shor's DLP quantum algorithm, they first replace the quantum Fourier transform A_q with A_{2^n} with $q \approx 2^n$, for the easy implementation purpose as follows.

$$|\Psi_1\rangle = |\mathcal{O}, \mathcal{O}, \mathcal{O}\rangle,$$

$$= \frac{1}{2^n} \sum_{a=0}^{2^{n-1}} \sum_{b=0}^{2^{n-1}} |a, b, \mathcal{O}\rangle$$

$$= \frac{1}{2^n} \sum_{a=0}^{2^{n-1}} \sum_{b=0}^{2^{n-1}} |a, b, aP + bQ\rangle$$

where

$$aP + bQ = \sum_i b_i P + \sum_i b_i Q$$

with

$$a = \sum_i a_i 2^i,$$

$$b = \sum_i b_i 2^i,$$

$$P_i = 2^i P,$$

$$Q_i = 2^i Q$$

can be performed efficient by classical Algorithm 4.5. However, in their implementation, Proos and Zalka have made some interesting modifications over Shor's original algorithm, as follows:

1. Eliminate the input registers $|a, b\rangle$. Only one accumulator register is needed for adding a fixed point P_i (wrt Q_i) to a superposition of points (called *group shift*) and two unitary transforms U_{P_i} and U_{Q_i} which acts on any basis state $|S\rangle$ representing a point on E are needed:

$$U_{P_i} : \ |S\rangle \to |S + P_i\rangle \ \ \text{and} \ \ U_{Q_i} : \ |S\rangle \to |S + Q_i\rangle.$$

2. Decompose the group shift. The ECDLP can be decomposed into a sequence of group shifts by constant classically known elements:

$$U_A : \ |S\rangle \to |S + A\rangle \ \ S, A \in E, \ \ A \text{ is fixed.}$$

In terms of the coordinators (x, y) of the points on E, the group shift is:

$$|S\rangle = |(x, y)| \ \ \to \ \ |S + A\rangle = |(x, y) + (\alpha, \beta)\rangle = ||(x', y')\rangle.$$

So the formulas for the group addition may be as follows:

$$\lambda \ = \ \frac{y - \beta}{x - \alpha} = \frac{y' + \beta}{x' - \alpha}, \ \ x' = \lambda^2 - (x + \alpha)$$

$$x, y \ \longleftrightarrow \ x, \lambda$$

$$\longleftrightarrow \ x', \lambda$$

$$\longleftrightarrow \ x', y'$$

$$x, y \ \longleftrightarrow \ x - \alpha, y - \beta$$

$$\longleftrightarrow \ x - \alpha, \lambda = \frac{y - \beta}{x - \alpha}$$

$$\longleftrightarrow \ x' - \alpha, \lambda = \frac{y' + \beta}{x' - \alpha}$$

$$\longleftrightarrow \ x' - \alpha, y' + \beta$$

$$\longleftrightarrow \ x', y'$$

where $\longleftrightarrow$ denotes the reversible operation.

3. Decompose the divisions. The divisions of the form $x, y \longleftrightarrow x, y/x$ may be decomposed into the following forms:

$$x, y \quad \xleftarrow{\text{Modular inverse}} \quad 1/x, y$$

$$\xleftarrow{\text{Multiplication}} \quad 1/x, y, y/x$$

$$\xleftarrow{\text{Multiplicative inverse}} \quad x, y, y/x$$

$$\xleftarrow{\text{Multiplication}} \quad x, 0, y/x.$$

4. Modular multiplication. The modular multiplication of the form $x, y \longleftrightarrow x, y, x \cdot y$ in

$$|x, y\rangle \to |x, y, x \cdot \mathrm{mod} p\rangle$$

may be decomposed into a sequence of modular additions and modular doublings as follows:

$$x \cdot y = \sum_{i=0}^{n-1} x_i 2^i y$$

$$\equiv x_0 y + 2(x_i y + 2(x_2 y + 2(x_3 y + \cdots))) \ (\mathrm{mod}\ p),$$

whereas the following series operations are performed in the third register:

$$A \quad \longleftrightarrow \quad 2A$$

$$\longleftrightarrow \quad 2A + x_i y \ (\mathrm{mod}\ p), \ i = n-1, n-2, \ldots, 0.$$

5. Modular inverse. The modular inverse is the most difficult operation in the quantum implementation. However, this can be done efficiently on classical computers by Euclid's algorithm. So, we suggest to use a classical computer rather than a quantum computer to solve the problem, making quantum and classical computations complimentary. Readers who are interested in the detailed quantum implementation of the modular inverse should consult [50] for more information.

Remark 4.4. The algorithm runs in time $\mathcal{O}(\lambda^3)$ and in space $\mathcal{O}(\lambda)$ using roughly 6λ qubits, where λ is the input length in bits.

Remark 4.5. One of the most important advantages of quantum algorithms for ECDLP over quantum IFP is that for breaking the same level of security cryptographic systems, namely, RSA and ECC, quantum algorithms for ECDLP use less quibit than that for IFP, as given in Table 4.3.

Table 4.3. Comparison between quantum IFP and ECDLP algorithms

Quantum IFP			Quantum ECDLP			Classical
λ	Qubits 2λ	Time $4\lambda^3$	λ	Qubits 7λ	Time $360\lambda^3$	Time
512	1024	$0.54 \cdot 10^9$	110	700	$0.5 \cdot 10^9$	c
1024	2048	$4.3 \cdot 10^9$	163	1000	$1.6 \cdot 10^9$	$c \cdot 10^8$
2048	4096	$34 \cdot 10^9$	224	1300	$4.0 \cdot 10^9$	$c \cdot 10^{17}$
3072	6144	$120 \cdot 10^9$	256	1500	$6.0 \cdot 10^9$	$c \cdot 10^{22}$
15360	30720	$1.5 \cdot 10^{13}$	512	2800	$50 \cdot 10^9$	$c \cdot 10^{60}$

Optimized Quantum Attack on ECDLP/ECC

As can be seen, Proos–Zalka algorithm [50] is only applicable to the ECDLP over finite field $\mathbb{F}_p$. However, in practice, elliptic curve cryptographic systems often use curves over the binary finite field $\mathbb{F}_{2^m}$. So later on, Kaye and Zalka [30] extended the Proos-Zalka algorithm applicable for $\mathbb{F}_{2^m}$. More specifically, they use the Euclid's algorithm for polynomials to compute inverses in $\mathbb{F}_{2^m}$.

Remarkably enough, Cheung et al. [9] proposed a quantum algorithm for attacking the ECDLP/ECC over $\mathbb{F}_{2^m}$ such as $\mathbb{F}_{2^{255}}$. More specifically, they improved an earlier algorithm by constructing an efficient quantum circuit (see, e.g., Fig. 4.5 for a particular example) element in binary finite fields and by representing elliptic curve points in projective coordinators. The depth of their circuit implementation is $\mathcal{O}(m^2)$, while the previous bound is $\mathcal{O}(m^3)$.

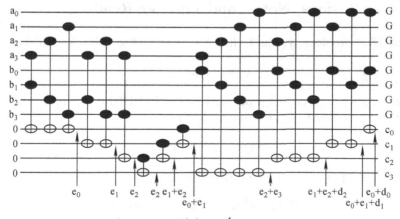

Figure 4.5. $\mathbb{F}_{2^4}$ multiplier with $P(x) = x^4 + x + 1$

Exercises and Problems for Sect. 4.3

1. Give a complete algorithmic description of the Kaye–Zalka's quantum ECDLP algorithm for $E(\mathbb{F}_{2^m})$.

2. Give a complete complexity analysis of Cheung et al.'s attack on ECDLP/ ECC over $E(\mathbb{F}_{2^m})$.

3. Design a quantum circuit to implement the Kaye–Zalka algorithm for breaking ECDLP/ECC in $E(\mathbb{F}_{2^m})$.

4. Van Meter and Itoh [64] developed a fast quantum modular exponentiation algorithm. Extend van Meter–Itoh's quantum modular exponentiation algorithm to fast quantum elliptic curve group operation.

5. Euclid's algorithm is suitable to compute gcd for both integers and polynomials, and more importantly, it can be performed in polynomial time even on a classical computer. What is the advantage to implement the quantum Euclid's algorithm?

6. The fastest known (classical) algorithm for solving the ECDLP in $F(\mathbb{F}_p)$ is Pollard's ρ method, runs in $\mathcal{O}(\sqrt{p})$ steps. As the periodicity lives at the very heart of the ρ method, it might (or should) be possible to implement a quantum version of the ρ method for ECDLP. Thus, give, if possible, a quantum implementation of the ρ algorithm for ECDLP.

4.4 Conclusions, Notes, and Further Reading

In the DLP problem, we aim to find the discrete logarithm k such that

$$y \equiv x^k \pmod{p},$$

where x, y, p are given and p prime, whereas in ECDLP, we aim to find the elliptic curve discrete logarithm k such that

$$Q \equiv kP \pmod{p},$$

where P is a point of order r on the elliptic curve

$$E\backslash\mathbb{F}_p : \ y^2 \equiv x^2 + ax + b \pmod{p},$$

$Q \in \langle P \rangle$, p is a prime. From a group-theoretic point of view, the computation of DLP is basically in the multiplicative group $\mathbb{Z}_p^*$, whereas the computation of ECDLP is mainly in the additive group $E(\mathbb{Z}_p)$. Compared to DLP, the computation of ECDLP is more difficult than that of DLP; the fastest general-purpose algorithm known for solving ECDLP is Pollard's ρ method, which has full-exponential expected running time of $\sqrt{\pi r}/2 = \mathcal{O}(\sqrt{p})$. As for the same level of security, the key length of DCDLP-based cryptography is shorter than that of IFP- or DLP-based cryptography. Thus, ECDLP-based cryptography is more useful in wireless security, where the key size is limited. However, this advantage of ECDLP-based cryptography is actually a serious disadvantage against the quantum attacks; as for the same level of security, ECC is easy to break than, e.g., RSA. In this chapter, same as the previous two chapters, the ECDLP problem and the classical solutions to the ECDLP problem are discussed, followed by an introduction to the ECDLP-based cryptographic systems. Finally, various quantum attacks on ECDLP- and ECDLP-based cryptographic systems are discussed.

The search for efficient classical solutions to ECDLP- and ECDLP-based cryptography and practical quantum attacks on ECDLP- and ECDLP-based cryptography is one of the most active ongoing research areas in mathematics, physics, computer science, and cryptography. Readers who wish to know more about ECDLP and methods for solving ECDLP are suggested to consult, e.g., [3–6, 10, 12, 15, 16, 18, 20, 26, 27, 32, 33, 41, 56, 57, 66]. In particular, the xedni calculus for ECDLP was proposed in [55] and analyzed in [24].

The security of elliptic curve cryptography (ECC) and the digital signature algorithm (ECDSA) are based on the infeasibility of the ECDLP. The idea to use elliptic curves, more specifically the ECDLP, as the basis to construct cryptographic systems was independently proposed by Miller [46] and Koblitz [31]. The following references provide more information on elliptic curves and elliptic curve (ECDLP based) cryptography: [1–4, 8, 10, 12, 13, 18–21, 23, 32, 33, 35, 37, 38, 42, 43, 45, 47, 48, 51, 52, 55–63, 65, 66, 69, 70].

Related literatures on quantum attacks on ECDLP- and ECDLP-based cryptography may be found in [7, 9, 14, 25, 29, 30, 49, 50, 53, 54, 67, 68].

For recent research progress on molecular DNA computation for ECDLP, readers are suggested to consult the following references and reference therein: [22, 28, 39].

REFERENCES

[1] G. Agnew, R. Mullin, S.A. Vanstone, An implementation of elliptic curve cryptosystems over $\mathbb{F}_{2^{155}}$. IEEE J. Sel. Areas Comm. **11**, 804–813 (1993)

[2] R.M. Avanzi, *Development of Curve Based Cryptography* (Ruhr-Universität Bochum, Germany, 2007), p. 12

[3] I. Blake, G. Seroussi, N. Smart, *Elliptic Curves in Cryptography* (Cambridge University Press, Cambridge, 1999)

[4]   I. Blake, G. Seroussi, N. Smart, *Advances in Elliptic Curves Cryptography* (Cambridge University Press, Cambridge, 2005)

[5]   J.W. Bos, M.E. Kaihara, T. Kleinjung et al., On the security of 1024-bit RSA and 160-bit elliptic curve cryptography, in *IACR Cryptology ePrint Archive* (2009), p. 19

[6]   J.W. Bos, M.E. Kaihara, T. Kleinjung et al., Solving a 112-bit prime elliptic curve discrete logarithm problem on game consoles using sloppy reduction. Int. J. Appl. Cryptography **2**(3), 212–228 (2012)

[7]   D.E. Browne, Efficient classical simulation of the quantum fourier transform. New J. Phys. **9**, 146, 1–7 (2007)

[8]   Certicom Research, *Certicom ECC Challenge*, 47 pp. (2009) http://www.certicom.com/index.php/the-certicom-ecc-challenge

[9]   D. Cheung, D. Maslo et al., On the design and optimization of a quantum polynomial-time attack on elliptic curve cryptography, in *Theory of Quantum Computation, Communication, and Cryptography Third Workshop, Theory of Quantum Computing 2008*. Lecture Notes in Computer Science, vol. 5106 (Springer, New York, 2008), pp. 96–104

[10] H. Cohen, G. Frey, *Handbook of Elliptic and Hyperelliptic Curve Cryptography* (CRC Press, Boca Raton, 2006)

[11] S. Cook, The P versus NP problem, in *The Millennium Prize Problems*, ed. by J. Carlson, A. Jaffe, A. Wiles (Clay Mathematics Institute/American Mathematical Society, Providence, 2006), pp. 87–104

[12] R. Crandall, C. Pomerance, *Prime Numbers – A Computational Perspective*, 2nd edn. (Springer, New York, 2005)

[13] N. Demytko, A new elliptic curve based analogue of RSA, in *Advances in Cryptology – EUROCRYPT 93*. Lecture Notes in Computer Science, vol. 765 (Springer, New York, 1994), pp. 40–49

[14] J. Eicher, Y. Opoku, *Using the Quantum Computer to Break Elliptic Curve Cryptosystems* (University of Richmond, Richmond, 1997), p. 28

[15] G. Frey, The arithmetic behind cryptography. Not. AMS **57**(3), 366–374 (2010)

[16] G. Frey, M. Müller, H.G. Rück, *The Tate pairing and the Discrete Logarithm Applied to Elliptic Curve Cryptosystems* (University of Essen, Germany, 1998), p. 5

[17] M.R. Garey, D.S. Johnson, *Computers and Intractability – A Guide to the Theory of NP-Completeness* (W.H. Freeman and Company, New York, 1979)

[18] D. Hankerson, A.J. Menezes, S. Vanstone, *Guide to Elliptic Curve Cryptography* (Springer, New York, 2004)

[19] G.H. Hardy, E.M. Wright, *An Introduction to Theory of Numbers*, 6th edn. (Oxford University Press, Oxford, 2008)

[20] J. Hoffstein, J. Pipher, J.H. Silverman, *An Introduction to Mathematical Cryptography* (Springer, New York, 2008)

[21] D. Husemöller, in *Elliptic Curves*. Graduate Texts in Mathematics, vol. 111 (Springer, New York, 1987)

[22] G. Iaccarino, T. Mazza, Fast parallel molecular algorithms for the elliptic curve logarithm problem over $GF(2^n)$, in *Proceedings of the 2009 Workshop on Bio-inspired Algorithms for Distributed Systems* (ACM, New York, 2008), pp. 95–104

[23] K. Ireland, M. Rosen, in *A Classical Introduction to Modern Number Theory*, 2nd edn. Graduate Texts in Mathematics, vol. 84 (Springer, New York, 1990)

[24] M.J. Jacobson, N. Koblitz, J.H. Silverman, A. Stein, E. Teske, Analysis of the Xedni calculus attack. Des. Codes Cryptography **20**, 41–64 (2000)

[25] R. Jain, Z. Ji et al., QIP = PSPACE. Comm. ACM **53**(9), 102–109 (2010)

[26] D. Johnson, A. Menezes, S. Vanstone, The elliptic curve digital signatures algorithm (ECDSA). Int. J. Inf. Sec. **1**(1), 36–63 (2001)

[27] O. Johnston, A Discrete Logarithm Attack on Elliptic Curves. IACR Cryptology ePrint Archive, vol. 575, p. 14 (2010)

[28] K. Karabina, A. Menezes, C. Pomerance, I.E. Shparlinski, On the asymptotic effectiveness of Weil descent attacks. J. Math. Cryptol. **4**(2), 175–191 (2010)

[29] P. Kaye, Techniques for Quantum Computing. Ph.D. Thesis, University of Waterloo, 2007, p. 151

[30] P. Kaye, C. Zalka, Optimized quantum implementation of elliptic curve arithmetic over binary fields. Quant. Inf. Comput. **5**(6), 474–491 (2006)

[31] N. Koblitz, Elliptic curve cryptography. Math. Comput. **48**, 203–209 (1987)

[32] N. Koblitz, in *A Course in Number Theory and Cryptography*, 2nd edn. Graduate Texts in Mathematics vol. 114 (Springer, New York, 1994)

[33] N. Koblitz, in *Algebraic Aspects of Cryptography*. Algorithms and Computation in Mathematics, vol. 3 (Springer, New York, 1998)

[34] N. Koblitz, Cryptography, in *Mathematics Unlimited – 2001 and Beyond*, ed. by B. Enguist, W. Schmid (Springer, New York, 2001), pp. 749–769

[35] N. Koblitz, A. Menezes, S.A. Vanstone, The state of elliptic curve cryptography. Des. Codes Cryptography **19**, 173–193 (2000)

[36] K. Koyama, U.M. Maurer, T. Okamoto, S.A. Vanstone, New public-key schemes based on elliptic curves over the ring $\mathbb{Z}_n$. (NTT Laboratories, Kyoto, 1991)

[37] K. Lauter, The advantages of elliptic curve cryptography for wireless security. IEEE Wirel. Comm. **11**(1), 62–67 (2004)

[38] H.W. Lenstra Jr., *Elliptic Curves and Number-Theoretic Algorithms* (Mathematisch Instituut, Universiteit van Amsterdam, Amsterdam, 1986)

[39] K. Li, S. Zou, J. Xv, Fast parallel molecular algorithms for DNA-based computation solving the elliptic curve logarithm problem over $GF(2^n)$. J. Biomed. Biotechnol. Article ID 518093, 10 (2008)

[40] A. Menezes, S.A. Vanstone, Elliptic curve cryptosystems and their implementation. J. Cryptol. **6**, 209–224 (1993)

[41] A. Menezes, T. Okamoto, S.A. Vanstone, Reducing elliptic curve logarithms in a finite field. IEEE Trans. Inf. Theor. **39**(5), 1639–1646 (1993)

[42] A. Menezes, P.C. van Oorschot, S.A. Vanstone, *Handbook of Applied Cryptography* (CRC Press, Boca Raton, 1996)

[43] A.J. Menezes, *Elliptic Curve Public Key Cryptography* (Kluwer, Dordrecht, 1993)

[44] J.F. Mestre, Formules Explicites et Minoration de Conducteurs de Variétés algébriques. Compositio Math. **58**, 209–232 (1986)

[45] B. Meyer, V. Müller, A public key cryptosystem based on elliptic curves over $\mathbb{Z}/n\mathbb{Z}$ equivalent to factoring, in *Advances in Cryptology, EUROCRYPT '96*. Proceedings, Lecture Notes in Computer Science, vol. 1070 (Springer, New York, 1996), pp. 49–59

[46] V. Miller, Uses of elliptic curves in cryptography, in *Lecture Notes in Computer Science*, vol. 218 (Springer, New York, 1986), pp. 417–426

[47] R.A. Mollin, *An Introduction to Cryptography*, 2nd edn. (Chapman & Hall/ CRC, London/West Palm Beach, 2006)

[48] R.A. Mollin, *Algebraic Number Theory*, 2nd edn. (Chapman & Hall/CRC, London/West Palm Beach, 2011)

[49] M.A. Nielson, I.L. Chuang, *Quantum Computation and Quantum Information*, 10th Anniversary edn. (Cambridge University Press, Cambridge, 2010)

[50] J. Proos, C. Zalka, Shor's discrete logarithm quantum algorithm for elliptic curves. Quant. Inf. Comput. **3**(4), 317–344 (2003)

[51] M. Rosing, *Implementing Elliptic Curve Cryptography* (Manning, New York, 1999)

[52] R. Schoof, Elliptic curves over finite fields and the computation of square roots mod p. Math. Comput. **44**, 483–494 (1985)

[53] P. Shor, Algorithms for quantum computation: discrete logarithms and factoring, in *Proceedings of 35th Annual Symposium on Foundations of Computer Science* (IEEE Computer Society, Silver Spring, 1994), pp. 124–134

[54] P. Shor, Polynomial-time algorithms for prime factorization and discrete logarithms on a quantum computer. SIAM J. Comput. **26**(5), 1484–1509 (1997)

[55] J.H. Silverman, The Xedni calculus and the elliptic curve discrete logarithm problem. Des. Codes Cryptography **20**, 5–40 (2000)

[56] J.H. Silverman, in *The Arithmetic of Elliptic Curves*. Graduate Texts in Mathematics, vol. 106, 2nd edn. (Springer, New York, 2010)

[57] J.H. Silverman, J. Suzuki, Elliptic curve discrete logarithms and the index calculus, in *Advances in Cryptology – ASIACRYPT '98*. Lecture Notes in Computer Science, vol. 1514 (Springer, New York, 1998), pp. 110–125

[58] N. Smart, *Cryptography: An Introduction* (McGraw-Hill, New York, 2003)

[59] M. Stamp, R.M. Low, *Applied Cryptanalysis* (Wiley, New York, 2007)

[60] A. Stanoyevitch, *Introduction to Cryptography* (CRC Press, West Palm Beach, 2011)

[61] D.R. Stinson, *Cryptography: Theory and Practice*, 2nd edn. (Chapman & Hall/CRC Press, London/West Palm Beach, 2002)

[62] H.C.A. van Tilborg, *Fundamentals of Cryptography* (Kluwer, Dordrecht, 1999)

[63] W. Trappe, L. Washington, *Introduction to Cryptography with Coding Theory*, 2nd edn. (Prentice-Hall, Englewood Cliffs, 2006)

[64] R. van Meter, K.M. Itoh, Fast quantum modular exponentiation. Phys. Rev. A **71**, 052320, 1–12 (2005)

[65] S.S. Wagstaff Jr., *Cryptanalysis of Number Theoretic Ciphers* (Chapman & Hall/CRC, London/West Palm Beach, 2002)

[66] L. Washington, *Elliptic Curves: Number Theory and Cryptography*, 2nd edn. (Chapman & Hall/CRC, London/West Palm Beach, 2008)

[67] C.P. Williams, *Explorations in Quantum Computation*, 2nd edn. (Springer, New York, 2011)

[68] C.P. Williams, S.H. Clearwater, *Ultimate Zero and One: Computing at the Quantum Frontier* (Copernicus, New York, 2000)

[69] S.Y. Yan, *Number Theory for Computing*, 2nd edn. (Springer, New York, 2002)

[70] S.Y. Yan, in *Primality Testing and Integer Factorization in Public-Key Cryptography*. Advances in Information Security, vol. 11, 2nd edn. (Springer, New York, 2009)

5. Quantum Resistant Cryptosystems

I think I can safely say that nobody understands quantum mechanics.

RICHARD FEYNMAN (1918–1988)
The 1965 Nobel Laureate in Physics

In this last chapter of the book, we shall introduce some cryptographic systems that resist all known quantum-computing attacks.

5.1 Quantum-Computing Attack Resistant

We have seen from previous three chapters that quantum computers, if can be built, can solve the famous infeasible IFP, DLP, and ECDLP problems efficiently in polynomial time, and more importantly, all the cryptographic systems and protocols, such as RSA, DHM, and ECC, based on these three types of infeasible problems can be broken in polynomial time. This might lead to the following wrong believing that quantum computers would speed up all computations and would solve all infeasible computational problems and break all cryptographic systems and protocols. It must be pointed out that quantum computers are not fast versions of classical computers, but just use a different and nonclassical paradigm for computation. They would speed up the computation for some problems such as IFP, DLP, and ECDLP by a large factor. However, for some other infeasible problems such as the famous traveling salesman problem and the shortest lattice problem, their computation power would just be the same as that of any classical computers. In fact, quantum computers have not been shown to solve any $\mathcal{NP}$-complete problems so far. Thus, the cryptographic systems and protocols based on some other infeasible problems, rather than IFP, DLP, and ECDLP, should be or may still be secure. That is, the quantum-computing based attacks would be invalid and no use for those cryptographic systems and protocols, whose

S.Y. Yan, *Quantum Attacks on Public-Key Cryptosystems*,
DOI 10.1007/978-1-4419-7722-9_5,
© Springer Science+Business Media, LLC 2013

security does not rely on the infeasibility of IFP, DLP, and ECDLP. More specifically, quantum computing is good at finding period, since fast Fourier transform (FFT) can be used to compute the period of a function, which is in turn can be extended to be a quantum Fourier transform (QFT) and can be run on a quantum computer. Thus, basically, quantum computers can speed up the computation for any periodic functions, as soon as FFT can be used. On the other hand, if the function is not periodic, or the computation is not suited for applying FFT, then the computation cannot generally be speed up by a quantum computer. As a consequence, any cryptographic systems and protocols, whose security does not rely on periodic problems or functions where FFT cannot be applied, should be potentially quantum-computing attack resistant.

Problems for Sect. 5.1

1. What is the main difference between a classical computer and a quantum computer?

2. In what sense or in which case a quantum computer can run fast than a classical computer?

3. Can you find a $\mathcal{NP}$-complete problem which can be solved by a quantum computer in polynomial time?

4. Explain why quantum computers are more powerful on solving the periodic functions. Justify your answer.

5. Explain why there must be many quantum-computing attack resistant cryptographic systems and protocols?

5.2 Coding-Based Cryptosystems

In this section, we introduce the most famous code-based cryptosystem, the McEliece system, invented by McEliece in [40]. One of the most important features of the McEliece system is that it has resisted cryptanalysis to date; it is even quantum computer resisted. The idea of the McEliece system is based on coding theory, and its security is based on the fact that decoding an arbitrary linear code is $\mathcal{NP}$-complete.

Algorithm 5.1 (McEliece's Code-Based Cryptography). Suppose
Bob wishes to send an encrypted message to Alice using Alice's public-key.
Alice generates her public-key and the corresponding private-key. Bob uses her
public-key to encrypt his message and sends it to Alice; Alice uses her own
private-key to decrypt Bob's message.

[1] **Key Generation:** Alice performs:

    [1-1]Choose integers k, n, t as common system parameters.

    [1-2]Choose a $k \times n$ generator matrix G for a binary (n, k)-linear code which
        can correct t errors and for which an efficient decoding algorithm exists.

    [1-3]Select a random $k \times k$ binary non-singular matrix S.

    [1-4]Select a random $k \times k$ permutation matrix P.

    [1-5]Compute the $k \times n$ matrix $\widehat{G} = SGP$.

    [1-6]Now $(\widehat{G}, t)$ is Alice's public-key whereas (S, G, P) is Alice's private-key.

[2] **Encryption:** Bob uses Alice's public-key to encrypt his message to Alice.
Bob performs:

    [2-1]Obtain Alice's authentic public key $(\widehat{G}, t)$.

    [2-2]Represent the message in binary string m of length k.

    [2-3]Choose a random binary error vector z of length n having at most t
        1's.

    [2-4]Compute the binary vector $c = m\widehat{G} + z$.

    [2-5]Send the ciphertext c to Alice.

[3] **Decryption:** Alice receives Bob's message m and uses her private-key to
recover c from m. Alice does performs:

    [3-1] Compute $\widehat{c} = cP^{-1}$, where P^{-1} is the inverse of the matrix P.

    [3-2] Use the decoding algorithm for the code generated by G to decode $\widehat{c}$
        to $\widehat{m}$.

    [3-3] Compute $m = \widehat{m}\widehat{S}^{-1}$. This m is thus the original plaintext.

Theorem 5.1 (Correctness of McEliece's Cryptosystem). In McEliece's cryptosystem, m can be correctly recovered from c.

Proof. Since

$$
\begin{aligned}
\widehat{c} &= cP^{-1} \\
&= (m\widehat{G} + z)P^{-1} \\
&= (mSGP + z)P^{-1} \\
&= (mS)G + zP^{-1}, \quad (zP^{-1} \text{ is a vector with at most } t \text{ 1's})
\end{aligned}
$$

the decoding algorithm for the code generated by G corrects $\hat{c}$ to $\hat{m} = mS$. Now applying S^{-1} to $\hat{m}$, we get $mSS^{-1} = m$, the required original plaintext.   □

Remark 5.1. The security of McEliece's cryptosystem is based on error-correcting codes, particularly the Goppa [39]; if the Goppa code is replaced by other error-correcting codes, the security will be severely weakened. The McEliece's cryptosystem has two main drawbacks:

(1) The public-key is very large.

(2) There is a message expansion by a factor of n/k.

It is suggested that the values for the system parameters should be $n = 1024$, $t = 50$, and $k \geqslant 644$. Thus, for these recommended values of system parameters, the public-key has about 2^{19} bits, and the message expansion is about 1.6. For these reasons, McEliece's cryptosystem receives little attention in practice. However, as McEliece's cryptosystem is the first probabilistic encryption and, more importantly, it has resisted all cryptanalysis including quantum cryptanalysis, it may be a good candidate to replace RSA in the post-quantum cryptography age.

Problems for Sect. 5.2

1. Compare the main parameters (such as encryption and decryption complexity, cryptographic resistance, easy to use, secret-key size, and public-key size) of RSA and McEliece systems.

2. Show that decoding a general algebraic code is $\mathcal{NP}$-complete.

3. Write an essay on all possible attacks for the McEliece coding-based cryptosystem.

5.3 Lattice-Based Cryptosystems

Cryptography based on ring properties and particularly lattice reduction is another promising direction for post-quantum cryptography, as lattice reduction is a reasonably well-studied hard problem that is currently not known to be solved in polynomial time or even subexponential time on a quantum computer. There are many types of cryptographic systems based on lattice reduction. In this section, we give a brief account of one of the lattice based on cryptographic systems, the NTRU encryption scheme. NTRU is rumored to stand for Nth-degree TRUncated polynomial ring, or Number Theorists aRe Us. It is a rather young cryptosystem, developed by Hoffstein, Pipher, and Silverman [26] in 1995. We give a brief introduction to NTRU; for more information, it can be found in [26, 27].

Algorithm 5.2 (NTRU Encryption Scheme). The NTRU encryption scheme works as follows.

[1] **Key Generation:**

[1-1] Randomly generate polynomials f and g in D_f and D_g, respectively, each of the form:

$$a(x) = a_0 + a_1 x + a_2 x^2 + \cdots + a_{N-2} x^{N-2} + a_{N-1} x^{N-1}.$$

[1-2] Invert f in $\mathcal{R}_p$ to obtain f_p, and check that g is invertible in f_q.

[1-3] The public-key is $h \equiv p \cdot g \cdot f_q \pmod{q}$. The private-key is the pair (f, f_p).

[2] **Encryption:**

[2-1] Randomly select a small polynomials r in D_r.

[2-2] Compute the ciphertext $c \equiv r \cdot h + m \pmod{q}$.

[3] **Decryption:**

[3-1] Compute $a = \text{center}(f \cdot c)$,

[3-2] Recover m from c by computing $m \equiv f_p \cdot a \pmod{q}$. This is true since

$$a \equiv p \cdot r \cdot \equiv +f \cdot m \pmod{q}.$$

In Table 5.1, we present some information comparing NTRU to RSA and McEliece.

Table 5.1. Comparison among NTRU, RSA and McEliece

	NTRU	RSA	McEliece
Encryption speed	N^2	$N^2 \approx N^3$	N^2
Decryption speed	N^2	N^3	N^2
Public-key	N	N	N^2
Secret-key	N	N	N^2
Message expansion	$\log_p q - 1$	$1 - 1$	$1 - 1.6$

Problems for Sect. 5.3

1. Give a critical analysis of the computational complexity of the NTRU cryptosystem.
2. NTRU is currently considered quantum resistant. Show that NTRU is indeed quantum resistant or may not be quantum resistant.
3. Lattice-based cryptography is considered to be quantum resistant. However, if designed not properly, it may be broken by traditional mathematical attacks without using any quantum techniques. For example, the Cai-Cusick lattice-based cryptosystem [17] was recently cracked completely by Pan and Deng [45]. Show that the Cai-Cusick lattice-based cryptosystem can be broken in Polynomial time by classical mathematical attacks.
4. It is widely considered that the multivariate public-key cryptosystems (MPKC, see [20]) are quantum resistant. As the usual approach to polynomial evaluation is FFT like, whereas quantum computation makes a good use of FFT to speed up the computation. With this regard, show that MPKC may not be quantum resistant.

5.4 Quantum Cryptosystems

It is evident that if a practical quantum computer is available, then all public-key cryptographic systems based on the difficulty of IFP, DLP, and ECDLP will be insecure. However, the cryptographic systems based on quantum mechanics will still be secure even if a quantum computer is available. In this section some basic ideas of quantum cryptography are introduced. More specifically, a quantum analog of the Diffie-Hellman key exchange/distribution system, proposed by Bennett and Brassard in [7], will be addressed.

First let us define four *polarizations* as follows:

$$\{0°, \ 45°, \ 90°, \ 135°\} \stackrel{\text{def}}{=} \{\rightarrow, \ \nearrow, \ \uparrow, \ \searrow\}. \tag{5.1}$$

The quantum system consists of a transmitter, a receiver, and a quantum channel through which polarized photons can be sent [8]. By the law of quantum mechanics, the receiver can either distinguish between the *rectilinear polarizations* $\{\rightarrow, \ \uparrow\}$, or reconfigure to discriminate between the diagonal polarizations $\{\nearrow, \ \searrow\}$, but in any case, he cannot distinguish both types. The system works in the following way:

1. Alice uses the transmitter to send Bob a sequence of photons, each of them should be in one of the four polarizations $\{\rightarrow, \ \nearrow, \ \uparrow, \ \searrow\}$. For instance, Alice could choose, at random, the following photons:

to be sent to Bob.

2. Bob then uses the receiver to measure the polarizations. For each photon received from Alice, Bob chooses, at random, the following type of measurements $\{+, \times\}$:

3. Bob records the result of his measurements but keeps it secret:

4. Bob publicly announces the type of measurements he made, and Alice tells him which measurements were of correct type:

5. Alice and Bob keep all cases in which Bob measured the correct type. These cases are then translated into bits $\{0, 1\}$ and thereby become the key:

6. Using this secret key formed by the quantum channel, Bob and Alice can now encrypt and send their ordinary messages via the classic public key channel.

An eavesdropper is free to try to measure the photons in the quantum channel, but, according to the law of quantum mechanics, he cannot in general do this without disturbing them, and hence, the key formed by the quantum channel is secure.

Problems for Sect. 5.4

1. Explain what are the main features of quantum cryptography?
2. Explain why the quantum key distribution is quantum-computing resistant?
3. Use the idea explained in this section to simulate the quantum key distribution and to generate a string of 56 characters for a DES key.
4. Use the idea explained in this section to simulate the quantum key distribution and to generate a stream of 128 or 256 characters for an AES key.

5.5 DNA Biological Cryptography

The world was shocked by a paper [1] of Adleman (the "A" in the RSA), who demonstrated that an instance of the NP-complete problem, more specifically, the Hamiltonian path problem (HPP), can be solved in polynomial time on a DNA biological computer (for more information on biological computing, see, e.g., [2] and [33]. The fundamental idea of DNA-based biological computation is that of a set of DNA strands. Since the set of DNA strands is usually kept in a test tube, the test tube is just a collection of pieces of DNA. In what follows, we shall first give a brief introduction to the DNA biological computation.

Definition 5.1. A *test tube* (or just tube for short) is a set of molecules of DNA (i.e., a multi-set of finite strings over the alphabet $\Sigma = \{A, C, G, T\}$). Given a tube, one can perform the following four elementary biological operations:

(1) **Separate** or **Extract**: Given a tube T and a string of symbols $S \in \Sigma$, produce two tubes $+(T, S)$ and $-(T, S)$, where $+(T, S)$ is all the molecules of DNA in T which contain the consecutive subsequence S and $-(T, S)$ is all of the molecules of DNA in T which do not contain the consecutive sequence S.

(2) **Merge**: Given tubes T_1, T_2, produce the multi-set union $\cup(T_1, T_2)$:

$$\cup (T_1, T_2) = T_1 \cup T_2 \tag{5.2}$$

(3) **Detect**: Given a tube T, output "yes" if T contains at least one DNA molecule (sequence) and output "no" if it contains none.

(4) **Amplify**: Given a tube T, produce two tubes $T'(T)$ and $T''(T)$ such that

$$T = T'(T) = T''(T). \tag{5.3}$$

Thus, we can replicate all the DNA molecules from the test tube.

These operations are then used to write "programs" which receive a tube as input and return either "yes" or "no" or a set of tubes.

Example 5.1. Consider the following program:

(1) Input(T)
(2) $T_1 = -(T, C)$
(3) $T_2 = -(T_1, G)$
(4) $T_3 = -(T_2, T)$
(5) Output(Detect(T_3))

The model defined above is an unrestricted one. We now present a restricted biological computation model:

Definition 5.2. A tube is a multi-set of aggregates over an alphabet Σ which is not necessarily $\{A, C, G, T\}$. (An aggregate is a subset of symbols over Σ.) Given a tube, there are three operations:

(1) **Separate**: Given a tube T and a symbol $s \in \Sigma$, produce two tubes $+(T, s)$ and $-(T, s)$, where $+(T, s)$ is all the aggregates of T which contain the symbols s and $-(T, s)$ is all of the aggregates of T which do not contain the symbol s.

(2) **Merge**: Given tube T_1, T_2, produce

$$\cup (T_1, T_2) = T_1 \cup T_2 \tag{5.4}$$

(3) **Detect**: Given a tube T, output "yes" if T contains at least one aggregate or output "no" if it contains none.

Example 5.2. (3-colourability problem) Given an n vertex graph G with edges $e_1, e_2, \cdots, e_z$, let

$$\Sigma = \{r_1, b_1, g_1, r_2, b_2, g_2, \cdots, r_n, b_n, g_n\}.$$

and consider the following restricted program on input

$$T = \{\alpha \mid \alpha \subseteq \Sigma,$$
$$\alpha = \{c_1, c_2, \cdots, c_n\},$$
$$[c_i = r_i \text{ or } c_i = b_i \text{ or } c_i = g_i], i = 1, 2, \cdots, n\}$$

(1) Input(T).

(2) for $k = 1$ to z. Let $e_k = \langle i, j \rangle$:

    (a) $T_{\text{red}} = +(T, r_i)$ and $T_{\text{blue or green}} = -(T, r_i)$.

    (b) $T_{\text{blue}} = +(T_{\text{blue or green}}, b_i)$ and $T_{\text{green}} = -(T_{\text{blue or green}}, b_i)$.

    (c) $T_{\text{red}}^{\text{good}} = -(T_{\text{red}}, r_j)$.

    (d) $T_{\text{blue}}^{\text{good}} = -(T_{\text{blue}}, b_j)$.

    (e) $T_{\text{green}}^{\text{good}} = -(T_{\text{green}}, g_j)$.

    (f) $T' = \cup(T_{\text{red}}^{\text{good}}, T_{\text{blue}}^{\text{good}})$.

    (g) $T = \cup(T_{\text{green}}^{\text{good}}, T')$.

(3) Output(Detect(T)).

Theorem 5.2. (Lipton, 1994) Any SAT problem in n variables and m clauses can be solved with at most $\mathcal{O}(m + 1)$ separations, $\mathcal{O}(m)$ merges, and one detection.

The above theorem implies that biological computation can be used to solve all problems in $\mathcal{NP}$, although it does not mean all instances of $\mathcal{NP}$ can be solved in a feasible way. From a computability point of view, neither the quantum computation model nor the biological computation model has more computational power than the Turing machine. Thus, we have an analogue of Church-Turing thesis for quantum and biological computations:

Quantum and Biological Computation Thesis: An arithmetic function is computable or a decision problem is decidable by a quantum computer or by a biological computer if and only if it is computable or decidable by a Turing machine.

This means that from a complexity point of view, both the quantum computation model and the biological computation model do have some more computational power than the Turing machine. More specifically, we have the following complexity results about quantum and biological computations:

1. Integer factorization and discrete logarithm problems are believed to be intractable in Turing machines; no efficient algorithms have been found for these two classical, number-theoretic problems; in fact, the best algorithms for these two problems have the worst-case complexity $\Theta\left((\log n)^2 (\log \log n)(\log \log \log n)\right)$. But, however, both of these two problems can be solved in polynomial time by quantum computers.

2. The famous Boolean formula satisfaction problem (SAT) and directed HPP are proved to be $\mathcal{NP}$-complete, but these problems, and in fact any other $\mathcal{NP}$-complete problems, can be solved in polynomial biological steps by biological computers.

Now we are in a position to discuss the DNA-based cryptography [23]. We first study a DNA analog of one-time pad (OTP) encryption; its idea may be described as follows:

1. **Plaintext encoding**: The plaintext: M is encoded in DNA strands.

2. **Key generation**: Assemble a large OTP in the form of DNA strands.

3. **OTP substitution**: Generate a table that randomly maps all possible strings of $M \to C$ such that there is a unique reverse mapping $M \leftarrow C$.

4. **Encryption**: Substitute each block of M with the ciphertext C given by the table to get $M \to C$.

5. **Decryption**: Reverse the substitutions to get $C \to M$.

The DNA implementation of the above scheme may be as follows:

1. **Plaintext in DNA**: Set one test tube of short DNA strands for M.

2. **Ciphertext in DNA**: Set another test tube of different short DNA strands for C.

3. **Key generation**: Assemble a large OTP in the form of DNA strands.

4. **OTP substitution**: Maps M to C in a random yet reversible way.

5. **Encryption : DNA substitution OTDs**: Use long DNA OTPs containing many segments; each contains a cipher word followed by a plaintext word. These word-pair DNA strands are used as a lookup table in conversion of plaintext into ciphertext for $M \to C$.

6. **Decryption**: Just do the opposite operation to the previous step for $C \to M$.

Just the same as stream cipher, we could use the operation XOR, denoted by $\oplus$, to implement the DNA OTP encryption as follows:

1. **DNA plaintext test tube**: Set one test tube of short DNA strands for M.

2. **DNA ciphertext test tube**: Set another test tube of different short DNA strands for C.

3. **Key generation**: Assemble a large OTP in the form of DNA strands.

4. **Encryption**: Perform $M \oplus$OTPs to get cipher strands; remove plaintext strands.

5. **Decryption**: Perform $C \oplus$ OTPs to get back plaintext strands.

Problems for Sect. 5.5

1. Explain how DNA computing can be used to solve the HPP.
2. Explain what are the main features of DNA biological cryptography?
3. Explain why DNA biological cryptography is a quantum-computing resistant?
4. DNA molecular biologic cryptography, e.g., Reif's OTP DNA cryptosystem developed in [23], is a new development in cryptography. Give a complete description and critical analysis of the Reif's DNA-based OTPs.
5. Write an assay to compare the main features of the classic, the quantum, and the DNA cryptography.

5.6 Conclusions, Notes, and Further Reading

Quantum-computing resistant, or quantum-attack resistant, or just quantum resistant cryptography is an important research direction in modern cryptography, since once a practical quantum computer can be build, all the public-key cryptography based on IFP, DLP, and ECDLP can be broken in polynomial time. As Bill Gates noted in his book [22]:

> We have to ensure that if any particular encryption technique proves fallible, there is a way to make an immediate transition to an alternative technique.

We need to have quantum resistant cryptographic systems ready at hand, so that we can use these cryptosystems to replace these quantum attackable cryptosystems. In this chapter, we only discussed some quantum resistant cryptographic systems, including quantum cryptography; interested readers should consult the following references for more information: [5, 6, 8, 9, 12, 13, 15, 18, 19, 21, 28–30, 34, 35, 37, 38, 42–44, 46, 52–54, 56, 57, 60, 61]. Note that in literatures, quantum-computing resistant cryptography is also called *post-quantum cryptography*. Springer publishes the proceedings of the post-quantum cryptography conferences [10, 16, 49, 62].

Just the same as quantum computing and quantum cryptography, DNA molecular computation is another type of promising computing paradigm and cryptographic scheme. Unlike the traditional computing model, DNA molecular computing is analog, not digital, so it opens a completely different phenomena to solve the hard computational problem. As can be seen from our above discussion, DNA computing has the potential to solve the NP-completeness problems such as the famous HPP and the satisfiability problem (SAT). Of course there is a long way to go to truly build up a practical DNA computer. Reader may consult the following references for more information on DNA computing and cryptography: [3, 4, 11, 14, 24, 25, 31, 36, 47, 48, 50, 55, 58].

Chaos-based cryptography [41, 51, 59] may be another type of good candidate for quantum resistant cryptography; readers are suggested to consult [32] for more information. Yet, there are another candidates for quantum resistant cryptography based on the conjectured difficulty of finding isogenies between supersingular elliptic curves [30], since the fastest known quantum algorithms for constructing isogenies between supersingular elliptic curves is exponential (however, the construction of isogenies between ordinary elliptic curves can be done in subexponential time).

REFERENCES

[1]   L.M. Adleman, Molecular computation of solutions to combinatorial problems. Science **266**, 1021–1024 (1994)

[2]   L.M. Adleman, On constructing a molecular computer, in *DNA Based Computers*, ed. by R. Lipton, E. Baum (American Mathematical Society, Providence, 1996), pp. 1–21

[3]   R.D. Barish, P. Rothemund, E. Winfree, Two computational primitives for algorithmic self-assembly: copying and counting. Nano Lett. **5**(12), 2586–2592 (2005)

[4]   Y. Benenson, B. Gill, U. Ben-Dor et al., An autonomous moleular computer for logical control of gene expressions. Nature **429**, 6990, 423–429 (2004)

[5]   C.H. Bennett, Quantum cryptography using any two nonorthogonal sates. Phys. Rev. Lett. **68**, 3121–3124 (1992)

[6] C.H. Bennett, Quantum information and computation. Phys. Today **48**(10), 24–30 (1995)

[7] C.H. Bennett, G. Brassard, Quantum cryptography: public key distribution and coin tossing, in *Proceedings of the IEEE International Conference on Computers Systems and Singnal Processing* (IEEE, New York, 1984), pp. 175–179

[8] C.H. Bennett, G. Brassard, A.K. Ekert, Quantum cryptography. Sci. Am. 26–33 (1992)

[9] E.R. Berlekampe, R.J. McEliece, H. van Tilburg, On the inherent intractability of certain coding problems. IEEE Trans. Inf. Theor. **IT-24**, 384–386 (1978)

[10] D.J. Bernstein, J. Buchmann, E. Dahmen (eds.), *Post-Quantum Cryptography* (Springer, Berlin, 2010)

[11] D. Boneh, C. Dunworth, R. Lipton et al., On the computational power of DNA. Discrete Appl. Math. **71**(1), 79–94 (1996)

[12] G. Brassard, Quantum computing: the end of classical cryptography? ACM SIGACT News **25**(3), 13–24 (1994)

[13] G. Brassard, C. Crépeau, 25 years of quantum cryptography. ACM SIGACT News **27**(4), 15–21 (1996)

[14] D. Bray, Pretein molecular as computational elements in living cells. Nature **376**, 6538, 307–312 (1995)

[15] D. Bruss, G. Erdélyi, T. Meyer, T. Riege, J. Rothe, Quantum cryptography: a survey. ACM Comput. Surv. **39**(2), Article 6, 1–27 (2007)

[16] J. Buchmann, J. Ding (eds.), in *Post-Quantum Cryptography*. Lecture Notes in Computer Science, vol. 5299 (Springer, Berlin, 2008)

[17] J.Y. Cai, T.W. Cusick, A lattice-based public-key cryptosystem. Inf. Comput. **151**(1–2), 17–31 (1999)

[18] E.F. Canteaut, N. Sendrier, Cryptanalysis of the original McEliece cryptosystem, in *Advances in Cryptology – AsiaCrypto'98*. Lecture Notes in Computer Science, vol. 1514 (Springer, Berlin, 1989), pp. 187–199

[19] P.-L. Cayrel, M. Meziani, Post-quantum cryptography: code-based signatures, in *Advances in Computer Science and Information Technology*. Lecture Notes in Computer Science, vol. 6059 (Springer, Berlin, 2010), pp. 82–99

[20] J. Ding, J.E. Gower, D.S. Schmidt, *Multivariate Public Key Cryptosystems* (Springer, Berlin, 2006)

[21] H. Dinh, C. Moore, A, Russell, McEliece and Niederreiter cryptosystems that resist quantum fourier sampling attacks, in *Advances in Cryptology – Crypto 2011*. Lecture Notes in Computer Science, vol. 6841 (Springer, Berlin, 2011), pp. 761–779

[22] B. Gates, *The Road Ahead* (Viking, New York, 1995)

[23] A. Gehani, T.H. LaBean, J.H. Reif, DNA-based cryptography, in *Molecular Computing*. Lecture Notes in Computer Science, vol. 2950 (Springer, Berlin, 2004), pp. 167–188

[24] T. Gramb, A. Bornholdt, M. Grob et al., *Non-Standard Computation* (Wiley-VCH, Weinheim, 1998)

[25] M. Guo, M. Ho, W.L. Chang, Fast parallel molecular solution to the dominating-set problem on massively parallel bio-computing. Parallel Comput. **30**, 1109–1125 (2004)

[26] J. Hoffstein, J. Pipher, J.H. Silverman, A ring-based public-key cryptosystem, in *Algorithmic Number Theory ANTS-III*. Lecture Notes in Computer Science, vol. 1423 (Springer, Berlin, 1998), pp. 267–288

[27] J. Hoffstein, N. Howgrave-Graham, J. Pipher, J.H. Silverman, W. Whyte, NTRUEncrypt and NTRUSign: efficient public key Algorithmd for a post-quantum world, in *Proceedings of the International Workshop on Post-Quantum Cryptography (PQCrypto 2006)*, (Springer, Berlin, 2006), pp. 71–77

[28] R.J. Hughes, Cryptography, quantum computation and trapped ions. Phil. Trans. R. Soc. Lond. Ser. A **356**, 1853–1868 (1998)

[29] H. Inamori, in *A Minimal Introduction to Quantum Key Distribution*. Centre for Quantum Computation, Clarendon Laboratory (Oxford University, Oxford, 1999)

[30] D. Jao, L. De Feo, Towards quantum-resistant cryptosystems from supersingular elliptic curve isogenies, in *Post-Quantum Cryptography*, ed. by Yang. Lecture Notes in Computer Science, vol. 7071 (Springer, Berlin, 2011), pp. 19–34

[31] N. Jonoska, G. Paun, G. Rozenberg (eds.), in *Molecular Computing*. Lecture Notes in Computer Science, vol. 2950 (Springer, Berlin, 2004)

[32] L. Kocarev, S. Lian, *Chaos-Based Cryptography* (Springer, Berlin, 2011)

[33] E. Lamm, R. Unger, *Biological Computation* (CRC Press, Boca Raton, 2011)

[34] A.K. Lenstra, H.W. Lenstra Jr., L. Lovász, Factoring polynomials with rational coefficients. Math. Ann. **261**, 515–534 (1982)

[35] H.W. Lenstra Jr., Lattices, in *Algorithmic Number Theory*, ed. by J.P. Buhler, P. Stevenhagen (Cambridge University Press, Cambridge, 2008), pp. 127–182

[36] R.Lipton, DNA solution of hard computational problems. Science **268**, 5210, 542–545 (1995)

[37] H.K. Lo, Quantum cryptography, in *Introduction to Quantum Computation and Information*, ed. by H.K. Lo, S. Popescu, T. Spiller (World Scientific, Singapore, 1998), pp. 76–119

[38] H. Lo, H. Chau, Unconditional security of quantum key distribution over arbitrary long distances. Science **283**, 2050–2056 (1999)

[39] F.J. MacWilliams, N.J.A. Sloana, *The Theory of Error Correcting Codes* (North-Holland, Amsterdam, 2001)

[40] R.J. McEliece, A Public-Key Cryptosystem based on Algebraic Coding Theory, pp. 583–584. JPL DSN Progress Report 42–44 (1978)

[41] I. MishkovskiK, L. Kocarev, Chaos-based public-key cryptography, in *Chaos-Based Cryptography*, ed. by L. Kocarev, S. Lian (Springer, Berlin, 2011), pp. 27–66

[42] P.Q. Nguyen, B. Vallée, *The LLL Algorithm: Survey and Applications* (Springer, Berlin, 2011)

[43] H. Niederreiter, Knapsack type cryptosystems and algebraic coding theory. Probl. Contr. Inf. Theor. **15**, 159–166 (1986)

[44] M.A. Nielson, I.L. Chuang, *Quantum Computation and Quantum Information*, 10th Anniversary edn. (Cambridge University Press, Cambridge, 2010)

[45] Y. Pan, Y. Deng, Cryptanalysis of the Cai-Cusick lattice-based public-key cryptosystem. IEEE Trans. Inf. Theor. **57**(3), 1780–1785 (2011)

[46] R.A. Perlner, D.A. Cooper, Quantum resistant public key cryptography, in *Proceedings of the 8th Symposium on Identity and Trust on the Internet*, Gaithersburg, MD, 14–16 April (ACM, New York, 2009), pp. 85–93

[47] C. Popovici, Aspects of DNA cryptography, Ann. Univ. Craiova, Math. Comput. Sci. Ser **37**(3), 147–151 (2010)

[48] J.H. Reif, Parallel biomolecular computation. Algorithmica **25**, 142–175 (1999)

[49] N. Sendrier (ed.), in *Post-Quantum Cryptography*. Lecture Notes in Computer Science, vol. 6061 (Springer, Berlin, 2010)

[50] H. Singh, K. Chugh, H. Dhaka, A.K. Verma, DNA-based cryptography: an approach to secure mobile networks. Int. J. Comp. Appl. **1**(19), 82–85 (2010)

[51] E. Solak, Cryptanalysis of chaotic ciphers, in *Chaos-Based Cryptography*, ed. by L. Kocarev, S. Lian (Springer, Berlin, 2011), pp. 227–254

[52] W. Trappe, L. Washington, *Introduction to Cryptography with Coding Theory*, 2nd edn. (Prentice-Hall, Englewood Cliffs, 2006)

[53] H. van Tilborg (ed.), *Encyclopedia of Cryptography and Security* (Springer, Berlin, 2005)

[54] H. van Tilburg, On the McEliece public-key cryptography, in *Advances in Cryptology – Crypto'88*. Lecture Notes in Computer Science, vol. 403 (Springer, Berlin, 1989), pp. 119–131

[55] R. Unger, J. Moult, Towards computing with protein. Proteine **63**, 53–64 (2006)

[56] J.L. Walker, *Codes and Curves* (American Mathematical Society and Institute for Advanced Study, Providence, 2000)

[57] C.P. Williams, *Explorations in Quantum Computation*, 2nd edn. (Springer, New York, 2011)

[58] E. Winfree, F. Liu, L.A. Wenzler et al., Design and self-assembly of two-dimensional DNA crystals. Nature **394**, 6693, 539–544 (1998)

[59] D. Xiao, X. Liao, S. Deng, Chaos-based Hash function, in *Chaos-Based Cryptography*, ed. by L. Kocarev, S. Lian (Springer, Berlin, 2011), pp. 137–204

[60] S.Y. Yan, *Cryptanalyic Attacks on RSA* (Springer, New York, 2009)

[61] S.Y. Yan, *Primality Testing and Integer Factorization in Public-Key Cryptography*, 2nd Edition (Springer, New York, 2010)

[62] B. Yang (ed.), in *Post-Quantum Cryptography*. Lecture Notes in Computer Science, vol. 7071 (Springer, New York, 2011)

Index

S.Y. Yan, *Quantum Attacks on Public-Key Cryptosystems*,
DOI 10.1007/978-1-4419-7722-9,
© Springer Science+Business Media, LLC 2013

About the Author

 Song Y. Yan received a Ph.D. in Number Theory in the Department of Mathematics at the University of York, England, and hold various posts at York, Cambridge, Aston, Coventry in the United Kingdom, and also various posts at MIT, Harvard, and Toronto in the North America. His research interests is mainly in Computational Number Theory, a inter-disciplinary subject of Number Theory, Computation Theory, and Mathematical Cryptography. He published, among others, the following five well-received research monographs and advanced textbooks in the field:

[1] *Perfect, Amicable and Sociable Numbers: A Computational Approach*, World Scientific, 1996.

[2] *Number Theory for Computing*, Springer, First Edition, 2000; Second Edition, 2002; Polish Translation, 2006 (Polish Scientific Publishers PWN, Warsaw); Chinese Translation, 2007 (Tsinghua University Press, Beijing).

[3] *Primality Testing and Integer Factorization in Public-Key Cryptography*, Springer, First Edition, 2004; Second Edition, 2009.

[4] *Cryptanalytic Attacks on RSA*, Springer, 2008. Russian Translation, 2010 (Russian Scientific Publishers, Moscow).

[5] *Computational Number Theory and Modern Cryptography*, Wiley, 2012.

S.Y. Yan, *Quantum Attacks on Public-Key Cryptosystems*,
DOI 10.1007/978-1-4419-7722-9,
© Springer Science+Business Media, LLC 2013

Printed in the United States
By Bookmasters